INDUSTRIAL CONFLICT AND DEMOCRACY

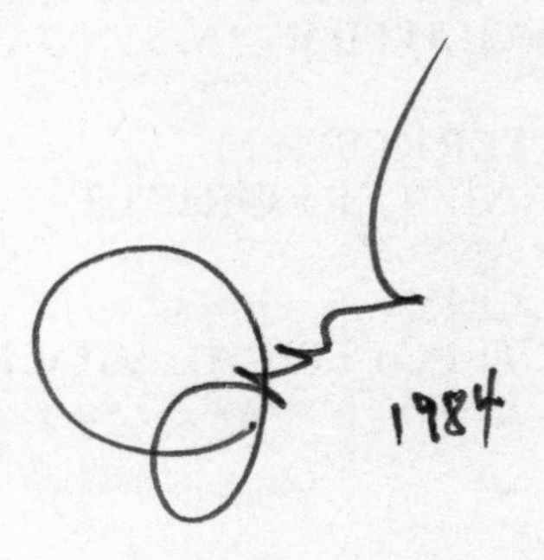

By the same author

ACROSS THE RIVER *(as Richard Jocelyn)*
THE LONG LONG WAR
PROTEST AND THE URBAN GUERRILLA
RIOT AND REVOLUTION IN SINGAPORE AND MALAYA
LIVING WITH TERRORISM
GUERRILLAS AND TERRORISTS
KIDNAP AND RANSOM
BRITAIN IN AGONY
THE MEDIA AND POLITICAL VIOLENCE

INDUSTRIAL CONFLICT AND DEMOCRACY

The Last Chance

Richard Clutterbuck

MACMILLAN

First Published 1984 by
THE MACMILLAN PRESS LTD
London and Basingstoke
Companies and representatives
throughout the world

Printed in Great Britain
by Camelot Press,
Southampton

British Library Cataloguing in Publication Data
Clutterbuck, Richard
Industrial conflict and democracy.
1. Industrial relations—Great Britain—History—20th century
I. Title
331'.0941 HD8391
ISBN 0-333-35805-8
ISBN 0-333-35806-6 Pbk

To Anne

Contents

List of Tables

List of Figures

Preface

In February 1982 I was asked by the Management Centre, Europe, to be Chairman of an international Top Management Forum in Stockholm to discuss the assumptions on which we in Europe should base our business planning for the next 5, 10 and 15 years. The two lead speakers were Kenneth Galbraith (USA) and Michael Shanks (UK) and there were other speakers from Italy, Switzerland, The Netherlands and the USA. Being Chairman involves both chairing the discussion and summing up so I had to concentrate harder than usual. I came away convinced that the world is on course for a boom in the 1990s but that, unless Britain can urgently overcome the internal stresses and strains which have dogged her industry since 1945 – the dreaded 'British disease' – she will miss out on this boom, with disastrous results, both economically and socially.

I was in 1982 already at work on researching the cost of industrial disputes – to the economy, to the taxpayer, to firms, to individual strikers and to the community at large. After the conference I decided to widen this research and try to produce some constructive ideas on how we can put our house in order before it is too late. That is what this book is about.

Much of the research was done by case studies and here we tried to get an even balance between management and trade unions. We never used quantitative questionnaires but went for interviews in depth and this meant that we asked a number of very busy managers, trade union officials, convenors and shop stewards to give up several hours of their time, often on several occasions. Of all the companies we approached, only one declined to accept a visit on the (not unreasonable) grounds that they were too busy. From the others, both management and union representatives, we received the most generous welcome and are immensely grateful. Inevitably their views often conflicted so we have, in the end, had to make our own judgements. For this reason, and because I think they would in any case prefer it, I have not named them personally.

For the first two years of my research Anne Ollivant was my research assistant but she was then suddenly struck down by illness so I was without her for the third year. This was a great blow both personally and professionally as she had established a remarkable rapport with many of the people we were interviewing, and this book, so much the result of her efforts, is dedicated to her. After she went I was extraordinarily lucky in that four of my students rallied round to put in many hours of part-time work to help me complete the book: Nick Baldwin, Amanda Bindon, Ann Condy and Jennie Wootton. Their views embraced a wide political spectrum but this never affected their enthusiasm and lively interest, both constructive and critical, which made them the most stimulating of comrades in arms and a constant inspiration. I also received unstinting help and support, as always, from our Politics Department secretaries, Fay Burgoyne and Sue Ridler. Including those in industry, I calculate that I owe about 100 people a great deal both for their time and for their encouragement in writing this book.

Exeter RICHARD CLUTTERBUCK
January 1984

List of Abbreviations

ACAS	Advisory, Conciliation and Arbitration Service
ACTSS	Association of Clerical, Technical and Supervisory Staff
AFL–CIO	American Federation of Labour–Congress of Industrial Organizations
APC	Atlantic Plant Corporation
APEX	Association of Professional Executive, Clerical and Computer Staff
ASLEF	Associated Society of Locomotive Engineers and Firemen
ASTMS	Association of Scientific, Technical and Managerial Staffs
AUEW	Amalgamated Union of Engineering Workers
BL	British Leyland
BLMC	British Leyland Motor Corporation
BMC	British Motor Corporation
BRB	British Railways Board
BSC	British Steel Corporation
CAD	Computer Aided Design
CBI	Confederation of British Industry
CDA	Cooperative Development Agency
CGT	General Confederation of Labour in France
COAB	Company Advisory Board
COHSE	Confederation of Health Service Employees
CNC	Computerized Numerical Control
CPF	Cooperative Productive Federation
CPRS	Central Policy Review Staff
CRS	Cooperative Retail Services Ltd.
CWS	Cooperative Wholesale Society
DGB	Deutscher Gewerksheftbund
DHSS	Department of Health and Social Security
DofE	Department of Employment

EEC	European Economic Community
EETPU	Electrical, Electronic, Telecommunications and Plumbing Union
FBU	Fire Brigades Union
FTATU	Furniture, Timber and Allied Trades Union
GDP	Gross Domestic Product
GEC	General Electric Company
ICOF	Industrial Common Ownership Finance
ICOM	Industrial Common Ownership Movement
IPD	International Property Development
ISTC	Iron and Steel Trades Confederation
JLP	John Lewis Partnership
JOL	Job Ownership Limited
JSSC	Joint Shop Stewards Committee
KME	Kirkby Manufacturing and Engineering Company Limited
LIFO	Last-in-first-out
LO	Federation of Labour in Sweden
M & S	Marks and Spencer
MDW	Measured Day Work
MIT	Ministry of International Trade and Industry in Japan
MOD	Ministry of Defence
NCB	National Coal Board
NEB	National Enterprise Board
NEC	National Executive Committee
NGA	National Graphical Association
NLRB	National Labour Relations Board in the US
NOP	Not on piecework
NSMM	National Society of Metal Mechanics
NUM	National Union of Mineworkers
NUPE	National Union of Public Employees
NUR	National Union of Railwaymen
ORC	Opinion Research Centre
PAYE	Pay as you earn
PLC	Public limited company
PLP	Parliamentary Labour Party

PSI	Policy Studies Institute
RCB	Regular Commissions Board
SAF	Swedish Employers Association
SDP	Social Democratic Party
SPD	Socialist Party of Germany
TASS	Technical, Administrative and Supervisory Section of the AUEW
TCO	Central Organization of Salaried Employees in Sweden
TGWU	Transport and General Workers' Union
TUC	Trades Union Congress
VDU	Visual Display Unit

Introduction

Industrial Britain has been losing ground steadily for years. Having had the highest living standards in Western Europe until a mere 50 years ago, we now have one of the lowest. Our productivity and growth rate are dismal and, most ominous of all, our capital investment per worker is one of the lowest in Europe, less than half that of Germany. In other words, the German worker has better machinery, in more efficient buildings with more efficient services and transport. The Germans are thus able to produce more efficiently, undersell us, make more profit and thereby invest more still and leave us still further behind. To take an analogy for a motor racing championship the Germans not only have a commanding lead; they have a better and faster car and, as they win more prize money, the disparity between the cars is programmed to become wider and wider.

On the face of it, this sounds like a terminal disease. Are we really caught in a vortex of decline leading inexorably to national bankruptcy? Unable to grow more than half the food we need, and having to import most of the raw materials for our factories we have to export or die. If we were to become wholly uncompetitive in the export market we would be starved of both food and work. The continuing real decline in standard of living would gather momentum, with its accompanying unrest. The International Monetary Fund would no doubt try to prop us up but, once its patience had eventually run out, we could be rescued only by an exponential growth in takeovers of our industry by foreign capital. Foreign companies, however would only invest – and maintain their investment – if they were confident that they could be assured of making a profit. They would thus demand cast iron agreements, with means of enforcement, for higher productivity at competitive labour costs per unit and, above all, some kind of guarantee (with, if need be, legal powers to enforce it) that production would not be disrupted. If this were effectively resisted by the trade unions, the foreign parent companies would either withhold or threaten to withdraw their investment (as US and French motor manufacturers have done) or demand even more stringent powers. Our people would be faced with a choice of near slavery or bankruptcy. The only worse fate would be an internal revolution. Such revolutions cause terrible suffering to ordinary families. As events in Poland and other Eastern European countries have proved, they lead to another form of near-slavery without averting bankruptcy – the worst of both worlds.

So are the British people doomed eventually to face a choice of being cheap labour for German and Japanese masters or of slumping into the shambles of revolution leading to the barren and hopeless fate of the Poles?

If all the graphs in the world moved in straight lines the answer would be yes; but luckily they do not and never have. The Germans and the Japanese have demonstrated that since 1945. The Japanese, in particular, have exposed what has for years been a precept of American capitalism: 'no war, strike or depression can destroy an established business or its profits as effectively as new and better methods or new and better equipment in the hands of an enlightened competitor'.

In the next 15 years, the revolution in robotics and information technology will have been consummated. A very different kind of industry will be in top gear in these countries which have used those 15 years well. They will be entering a boom period and their peoples will have a much better lifestyle. They will probably be working a 30 hour week and few, if any, of their workers will be doing repetitive work on assembly-lines. More important, the unit production cost of most products will be lower in real terms – i.e. the workers in these countries will need to work fewer hours to earn more, and their economies will be able to maintain the service and leisure industries to meet the needs of the new lifestyle.

It is sad, though not surprising, that this change is viewed with alarm in many quarters, in the same way that the Luddites regarded the first industrial revolution with alarm in 1811. Yet, few would now contest that the first industrial revolution has in the end created a far higher standard of living for a far larger number of people working far shorter hours. Certain skills which were relevant in 1760 had ceased to be relevant by 1811 and those who had the sense to do so learned new skills. The same will apply in the coming ten years. There will be painful changes but, in the end, there will be a higher standard of living with fewer hours work and, above all, a release from the assembly line.

If we fail to grasp the opportunities offered by this technological revolution we shall fail to compete in terms of unit cost with other industrial countries, and will face one or other of the fates described above. Therein lies the challenge. On the other hand, we can, if we choose, accept the challenge and compete. There is a popular fallacy that the Japanese are years ahead in robotics and information technology. They are not – though they *or we* could still make a technological breakthrough. At present, however, all the Japanese are doing is adapting their industry to the existing technology while British firms are doing so more slowly or not at all. There is time for us to catch up in these 15 years if we turn our minds to it. Therein lies the unrepeatable opportunity. In 15 years time, most German and Japanese factory plant, like our own, will be out of date and they will have replaced it. There is no reason why we should not at one bound be as up to date as they will be.

To catch up in one bound we need two things: a much faster capital investment and a readiness on the part of management and labour to adapt to new techniques. The more of this capital we can generate ourselves the better, but some will inevitably have to come from foreign investment. That investment will only flow if we can convince the investors that we are able and willing to adapt and that our industry has rid itself of continuous disruption and restrictive practices – in other words that we have cured ourselves of the British disease.

Our decline and the German and Japanese advance can be compared to that of three teams in the Football League. Teams can drop within a very few seasons from the 1st to the 3rd or 4th Divisions. Others, however, can climb just as quickly to the first. The UK was top of the 1st Division in the European economic league in 1945, and running with the USA, Canada and a few others in the top slots of the world league. Germany and Japan started on the floor and are now up with the leaders and (in Japan's case) still rising fast.

They would not, however, have achieved this unless their whole team had been playing to win. No football team could hope to win if, say, the backs were kicking through their own goal to spite the manager or because they resented the goalkeeper being paid more than they were; or if the directors had so little confidence that the team would play to win that they invested neither in players nor in development of the ground. That is the heart of the British disease.

We can look at the structure of German and Japanese industry and, in particular, at their techniques for worker-participation at shop-floor and boardroom levels, and for industrial relations. We can also, however, look nearer home. Some of our own firms have solved their problems as well as any overseas – e.g. the John Lewis Partnership and Marks and Spencer. Both will be examined more fully later and so will other successful firms in UK and overseas. Lessons will also be drawn from some of the bad experiences and manifestations of the British disease in action.

The irony is that managements, trade unions, investment institutions and governments are aware of the disease and know well enough the lines on which it must be cured. All, however, are more fearful of prejudicing their own bargaining power, interests or aspirations than about failure to cure the disease. Their fears are real enough in the short term, but taking too much counsel of them in the short term will exact a disastrous price in the long term.

Managements know, for example, that the cooperation of their workforce in joint efforts to secure higher productivity and avoid self-damaging disruption will only be achieved if there is greater worker-participation in management, so that the workers become convinced and aware of the cost to themselves of poor or disrupted production. Yet managements fear that workers on the board will inhibit management decisions; that meetings will be prolonged and argumentative; that they will provide trade union officials with inside information which they can and will use against management in wage negotiations; and that there will be leaks of commercial secrets to competitors. The Germans have largely resolved this problem, as will be discussed, by the use of separate supervisory and executive boards, and by direct election for worker directors by secret ballot on the shop floor, but in Britain there are still a lot of management anxieties to be set at rest.

Many trade unionists fear the robotic revolution, concerned about the manpower redeployment it involves, without seeing the promise it offers of rescuing human beings from repetitive assembly line work and eventually of earning more per head with fewer hours work. Regarding worker participation they are anxious lest directly elected worker directors may erode their position with the work force. They are also reluctant (as was shown in the reaction to the Bullock Report) to join Boards of Management themselves, since they would then find it difficult to face management in collective bargaining over decisions to which they have been a party (even if in a minority) on the Board. Some trade unionists may be 'industrial militants' (not necessarily political militants) who believe that they can only hold the loyalty of their members by being always seen as conducting tough negotiations to win more for them than the management would otherwise pay, and who therefore see confrontation as their *raison d'être*.

Investment institutions (e.g. insurance companies, merchant banks, trusts and pension fund trustees) perceive that their prime responsibility is to their clients who benefit from the earnings and growth of their investment. If investment in, say, motor manufacture in Germany or Japan will yield higher dividends and growth than investment in Britain, they feel it their duty to invest overseas, even though in the long term this adds to the decline of British industry and thus, eventually, of their own economic base.

This is where governments can exert most influence, by the judicious application of incentives and controls which will make it *more* profitable to invest in UK, both for internal and foreign investors. This is part of the secret of Lee Kuan Yew's success in Singapore. Yet British governments over the decades of decline have taken greater counsel of their short-term fears of an adverse balance of payments and of applying unpopular measures which will lose votes at the next General Election.

Home investment and confidence are vital for our future. We must attract both domestic and foreign money into industry. To do this we must find ways of overcoming these fears and inhibitions so that all members of the team are attacking our competitors' goal, not kicking through our own.

There are formidable but not unsurmountable difficulties and prejudices to be overcome before we can achieve this but it can be achieved, and it must be achieved if we are to seize this once-for-all-time chance to catch the tide of the microelectronic revolution and compete on equal terms in the boom of the 1990s.

That is the theme of this book. Part I looks at Britain's place in the world, past, present and future; at our present industrial relations organization; at the ideas of our various governments in the past for improving our performance; and at some of the approaches of other countries which seem to do better than we do.

Part II assesses the price we pay for our industrial conflict, nationally, collectively and individually, as illustrated by case studies.

PartIII examines some case studies of successful approaches to communication and participation, in Britain and elsewhere.

Part IV considers some of the ways in which we can tackle the British disease in the future and what Britain could be like in the 1990s if we succeed.

Part I
The Problem

I Britain's Place in the World

'YOU CAN GO ON TO TOMORROW'

'You cannot go back to yesterday but you can go on to tomorrow.' Maureen Martin, a policewoman shot in the back by a gunman and paralysed, described this as 'the best bit of advice I ever had'.[1] Britain could well take that advice too. For more than a century she led the industrial world or at least contested the lead with the USA and Germany. Being a small country with few natural resources, she was throughout that period able to feed only half her population so she had to live by exporting manufactured goods for which she had to import 80 per cent of the raw materials. Her industry had to be competitive and to supplement her industrial exports she built up invisible exports, by using her wealth to provide banking and insurance services and to invest overseas. The dividends from her overseas investments, above all, were in effect paid in the form of food and raw materials, beef from Argentina, butter from New Zealand and rubber and tin from Malaya.

Britain, alone among all the nations of the world, fought in both world wars from start to finish. Millions of her people were diverted from manufacturing goods for export to fight or to make munitions. By 1943–44, the volume of British exports had fallen to one-fifth of its 1938 level; but she still had to import food and raw materials. Her massive overseas investments could not earn enough in interest to fill this reduction in exports. Many of the overseas capital investments had therefore to be sold, though they have picked up again since Mrs Thatcher's Government released currency controls in 1979.

One of the results of the war was that from 1945 onwards Britain had to make up for the interest and earnings from these lost overseas investments by even more manufactured exports. With factories still dislocated and starved of money to replace ageing plant, British industry was in no state to recapture its markets and still less to build new ones. On the other hand, Germany, generally re-equipped with new industrial plant by the USA under the Marshall Plan, was able to build its exports to unprecedented heights. Britain's share of world trade, so far from increasing to make up for the lost earnings from overseas, steadily declined.

To put this into figures: in 1938 Britain's share of world trade had been 12.1 per cent. In 1948, despite the war, her share had still held up reasonably well at 11.5 per cent, but this was largely because Germany (1.4 per cent) and Japan (0.4 per cent) had as yet scarcely reentered the race. Thereafter Britain lost ground steadily as the volume of world trade grew.

TABLE 1.1 *Share of world trade (general)*[2]

	1948 %	*1963* %	*1975* %
Britain	11.5	7.9	5.0
Germany	1.4	9.5	10.3
Japan	1.4	3.5	6.4

HOW BRITAIN WAS OVERTAKEN BY GERMANY AND JAPAN

There was, however, another much more significant reason for Britain losing ground to Germany and Japan. The British people relapsed into a somewhat resentful feeling that they had borne the brunt of the war on everyone else's behalf; that they had won it only after five years of intense effort and self-denial, and they had a right to relax a little. They also had expectations of a fairer and more egalitarian post-war Britain. The 1945 election, despite Churchill's personal popularity, saw the biggest swing to the left in British Parliamentary history. The analysis of postal votes from overseas theatres of war suggested that it was the soldiers who mainly swung the vote for Labour; they above all wanted a new deal. Nevertheless, pay differentials between management and labour persisted, and although these differentials soon became far smaller than in Germany, many felt inclined to use the strike weapon to press their claims, and some of this attitude persisted after Churchill was re-elected in 1951.

Not so in Germany and Japan. Both nations had been humiliated and materially smashed to a degree unknown in previous wars. In particular, in the winters of 1945–46 and 1946–47 (a bitterly cold one) many Germans tasted almost unimaginable misery, living in cellars under the rubble of their ruined houses, short of fuel, with food rationed to subsistence level. They were proud and practical people and had an intense desire to rebuild a civilized life for themselves and their fatherland – and perhaps most of all to show themselves and the world that, despite being overwhelmed by massive superiority in numbers and armaments, a German was still worth two of any of those who had overwhelmed them. With that spirit, which carried them right through the 1950s, 1960s, and 1970s, the last thing they wanted – or tolerated from their comrades in the team – was to kick through their own goal. They made good use of Marshall Aid to reconstruct their industry. Guided (ironically) by a delegation of British trade unionists, they designed an industrial relations system geared to increasing production by the team rather than to competition between rival members of it for shares of the reward. This industrial relations system will be examined more fully in Chapter 4 so it suffices to say now that, in the period 1953–75 Britain lost 24 times more working days per thousand in industrial disputes than Germany did.[3]

Less tangible, but even more impressive, was the attitude towards work. One example provides a microcosm of this. In 1951 the army of occupation required a building close to a barracks newly taken over and under the Occupation Statute. The Burgomeister had to requisition a road house on the main Hanover–Bremen road. The owner did a thriving trade and his parking lot across the road was always crowded with cars and trucks. He appealed against the requisition but there was no other suitable building in the vicinity so, after a court hearing, the Burgomeister had no option but to order compliance. The owner finally bowed to this decision one week before the deadline by which he had to vacate the restaurant. Next morning, men were seen digging in the field behind the parking lot. They were putting in new foundations, electric cables and piping for gas, water and drainage. Seven days after the first turf was cut, there was a new restaurant, complete with kitchen, hot and cold water, heating, electric light, and cloths on the tables, ready to serve the customers. The day the old restaurant was vacated, the parking lot was as full of cars and trucks as ever and the customers were simply directed to the new restaurant in the field behind instead of the old one across the road.[4]

This dynamic and determined attitude characterized the reconstruction of Germany; it also permeated German industry though it took little time for the retooling of German factories to catch up. In 1953, British Gross Domestic Product (GDP) per head was nearly 2½ times that of Germany. The growth rate, however, was the other way round – Germany's being more than 2½ times greater than Britain's, so the writing was on the wall. By 1965 German GDP in real terms had more than doubled; by 1970 it had drawn level and by 1975 was over 1½ times that of Britain; this was matched by real earnings, the German worker earning 1½ times more; but because their output per man was higher and their prices therefore more competitive, German exports were double those from Britain; and, because efficient production creates profit and profit provides for reinvestment, the capital investment per worker in Germany by 1975 became the highest in the world, higher than that in the USA or Japan and double that in Britain. To our disgrace, capital investment per worker in Britain in 1975 was lower even than that in Italy.[5]

ADJUSTING TO TECHNOLOGICAL CHANGE

In the late 18th Century, 92 per cent of the British people lived or worked on the farms. The second biggest profession was domestic service. Today, 2 per cent work on the farms and a negligible proportion work in domestic service. During 200 years we have adjusted to this quite revolutionary change in our pattern of employment.

Even more dramatic, but seldom recognized, has been the reduction *in the last ten years* of the number employed in manufacturing industry. In 1972, 42 per cent of the population were employed in manufacturing. In 1982 it had fallen to 29 per cent. We have failed to adjust and we have over 3 million unemployed.

Projecting that trend ahead, and allowing for automation (whether we are in the lead or tailing along behind) that 29 per cent may have fallen to 10 per cent by the end of the century.[6]

Another cause of our high unemployment is our failure to adjust our working hours to the automation which has been happening over the past century, brought about by mass production techniques and machine tools. In the 1860s the average working week was 65 hours. By 1939 it was between 40 and 45. Since the Second World War, however, the working hours have fallen only marginally – though they are more commonly worked with overtime in a 5-day week rather than in a 5½ day week. If we had adjusted to the opportunities made available to us by technological advances during this period, the average working week should already be down to 32 hours a week – with no reduction at all in production[7]. The job-sharing and greater prosperity this should have brought would have resulted in our being able to use – and afford – higher employment in service and leisure industries. But Britain – especially in the past 25 years – has signally failed to adjust.

Why is this? The trade unions are well aware of the problem and have repeatedly called for a shorter working week; but with one glaringly obvious fallacy – they demand that production workers should have same or higher pay for the shorter working week *without increasing production*.

One factor in this is that we are slow to adjust to the idea of shorter working hours. It is as if there were some immutable human norm (like 3 meals a day or 8 hours sleep) that we should spend 40 of the 168 hours in the week at work. We are not yet ready for more leisure or the idea of producing more in fewer hours.

There has been an element of insincerity in some of the trade union demands for a shorter working week. These have in many cases been demands for a shorter *basic* working week, so that more of the hours actually worked are paid at overtime rates. There has been little shop floor pressure for working shorter hours as such, but simply for a way of being paid more for the same number of hours, without any corresponding increase in production. This is a human enough desire, but can clearly only result in higher unit costs and therefore a lower share of the market against competitors both in the home and export markets – resulting in more unemployment. Government figures given in Parliament on 21st October 1982 indicated that, if British manufacturers still held the same share of the home market as in 1970, 1½ million out of the 3¼ million unemployed would still have jobs.

This is another manifestation of a sadly unattractive change in our national character over the past century. The 19th century friendly societies and trade unions developed to provide an organization whereby those in work could help those out of work (through sickness or factory closures or lockouts or strikes). This attitude was still recognizable in the behaviour of coal miners between 1960 and 1970, when they accepted the joint persuasion of Alf Robens (NCB) and Will Paynter (NUM) to restrain their wage demands in order that the older pits in South Wales and the North East would not be made economically unviable, thereby throwing whole mining communities (with no other local employment available) onto the dole. This attitude has changed, and not only in the mining industry. Many of those in work press for higher real earnings without increase in production well knowing that this will mean a lower volume of production and thence more unemployment for others. This may be an inevitable by-product of the welfare state but it is a sad one. Those in work have never earned more on average in real terms, and the lower *national* standard of living is due to the average being pulled down by the 3¼ million out of work. This is another version of 'two nations' and probably contributes to the unpopularity of the trade unions, which are seen as the driving force of the demand for more for the haves at the expense of the have-nots. Extra earnings from technological advance have been used to widen this gap whereas they should have been used to spread the benefits by lowering unit production costs and sharing the work. Both compassion and solidarity have been eroded.

The continuous pressure for more pay for less production has resulted in the anomaly that, for each family of four, we are as a nation spending £1000 per year more than we are creating.[8] A high percentage of this has been spent in subsidizing uneconomic production in publicly-owned industries. During the middle 1970s, much of this shortfall was made up by massive borrowing against future production of North Sea Oil. As a result, a lot of the national wealth created by North Sea Oil is being poured into the vacuum of past debts and propping up the disparity between what we earn and what we produce instead of being invested in modernizing industry to become more competitive. Since North Sea Oil is a transient asset, we are thus aborting our technological development and thereby in effect depriving our children and grandchildren by consuming their seed corn.

THE CYCLE OF INDUSTRIAL REVOLUTIONS

The most pernicious myth prevalent in British industry is that technological development must result in fewer jobs and lower earnings. This is tantamount to saying that our agriculture would be viable if we still tilled the soil by hand instead of using the plough, or that our textile industry would thrive if weaving were still done by hand instead of by machine, as the Luddites tried to insist in 1811. We could never have fed our population or provided work for our growing workforce without the industrial revolution. With very rare and very small dips in the rising curve, average real earnings and standards of living for manual workers have risen continuously and often dramatically over the past 200 years. In particular, they have more than doubled in the past 35 years.

There seems to be a cycle of about 50 years in the pattern of industrial revolutions, each divided into 25 years of depression coinciding with the development of new techniques followed by 25 years of boom when this new technology has put a modernized industry into top gear. Such booms seem to run out of steam in about 25 years, at which a further 25 year depression/development phase begins, leading on to its own 25 year boom. Thus, starting with the development of railways and projecting ahead to the culmination of the current microelectronic revolution, this is the pattern:

1825–50	Depression	Development of railways.
1850–75	Boom	The railway age.
1875–1900	Depression	Development of mass production of steel, electricity and chemicals.
1900–25	Boom	Mass production age, including diversion to munitions for World War One.
1925–50	Depression	Development of air transport and electronics, accelerated by the Second World War.
1950–70	Boom	Mass transport and automation. The 'consumer society' with wide distribution of consumer durables.
1970–95?	Depression	Development of microelectronics, robotics, information technology and electric power by nuclear fusion.
1995–?	Boom	Abundant production of low cost consumer goods with greater wealth and leisure for a high standard of living, and no energy shortage in industrial countries.

This second and more affluent consumer society may not suit everyone's taste but, as with other periods of affluence, most of those who do not share in its benefits will envy those who do. Virtually all jobs will need thought and decision with little repetitive process work. Closed circuit cable television will enable many more people to work at home, with instant access to information and face to face discussion on the screens. Household chores will be greatly reduced and many people will earn all they want in say, 3½ working days (or four shifts) per week. Many more will work in service or leisure industries, and more will have time to use them. More travel will be for leisure rather than commuting to work or to attend conferences. There will probably also be a thriving 'moonlight' economy giving scope to those who prefer less leisure now to earn more for more ambitious holidays at intervals.

There will be many problems. Some people, as in every age, may fail to adjust to this new society and the contrast between the affluent majority and the deprived or drop-out minority may be even starker than before. This may apply even more to international comparisons resulting not only in a greater gap between rich and poor nations but also greater inequities between rich and poor in the poorer nations. There is no way in which all the world can consummate a microelectronic revolution by 1995.

The USA, Japan, Switzerland, Sweden, Germany and many other members of the EEC assuredly will. If Britain fails to do so her decline and consequent disgruntlement will gather momentum. But she is in no state to compete in this race so long as she is debilitated by the British disease. So the first essential is to tackle that disease.

In the next few chapters we will look at the organization for industrial relations in Britain and the record of government involvement in them; then at some of the approaches in foreign countries as a prelude to examining the price we pay for industrial conflict in Britain.

2 The British Industrial Relations Structure

MANAGEMENT ORGANIZATION FOR INDUSTRIAL RELATIONS

The industrial relations structure in Britain is one of the most complicated in the world, a heritage of having pioneered the industrial revolution and the development of trade unions. In contrast with Germany, where there are normally only one or two unions bargaining for the workforce in any one factory, British companies are saddled with a plethora of rival unions, whose members overlap in almost every operation; and of levels and channels of negotiation, providing unlimited opportunities for misunderstanding or confounding of agreements reached in good faith, for unions playing off different levels of management and for management playing off different unions against each other.

The organization of the management structure and the machinery for industrial relations vary considerably from company to company. They depend upon the size of the firm and the specific roles or tasks attributed to each of the individuals and levels within the management structure. Figure 2.1 gives an example of the management 'family tree' of a large industrial firm. In smaller firms many of the functions would be combined and the number of levels reduced.

A Board of Directors may include a number of part-time, non-executive directors bringing a particular expertise (e.g. specialist technical knowledge or contact with an associated industry) and a Chairman who may also be Managing Director. Apart from these, every other director is responsible for a particular area of operations, as is usually clear from his title. Responsibility for industrial relations, however, is by no means so clear. Formally, it lies with the Personnel Director and those under him, particularly the Industrial Relations Manager. It is their function to provide practical advice and guidance, and the necessary services to all levels of management. Specifically the Personnel Director, who deals with a range of activities shown in Figure 2.1,

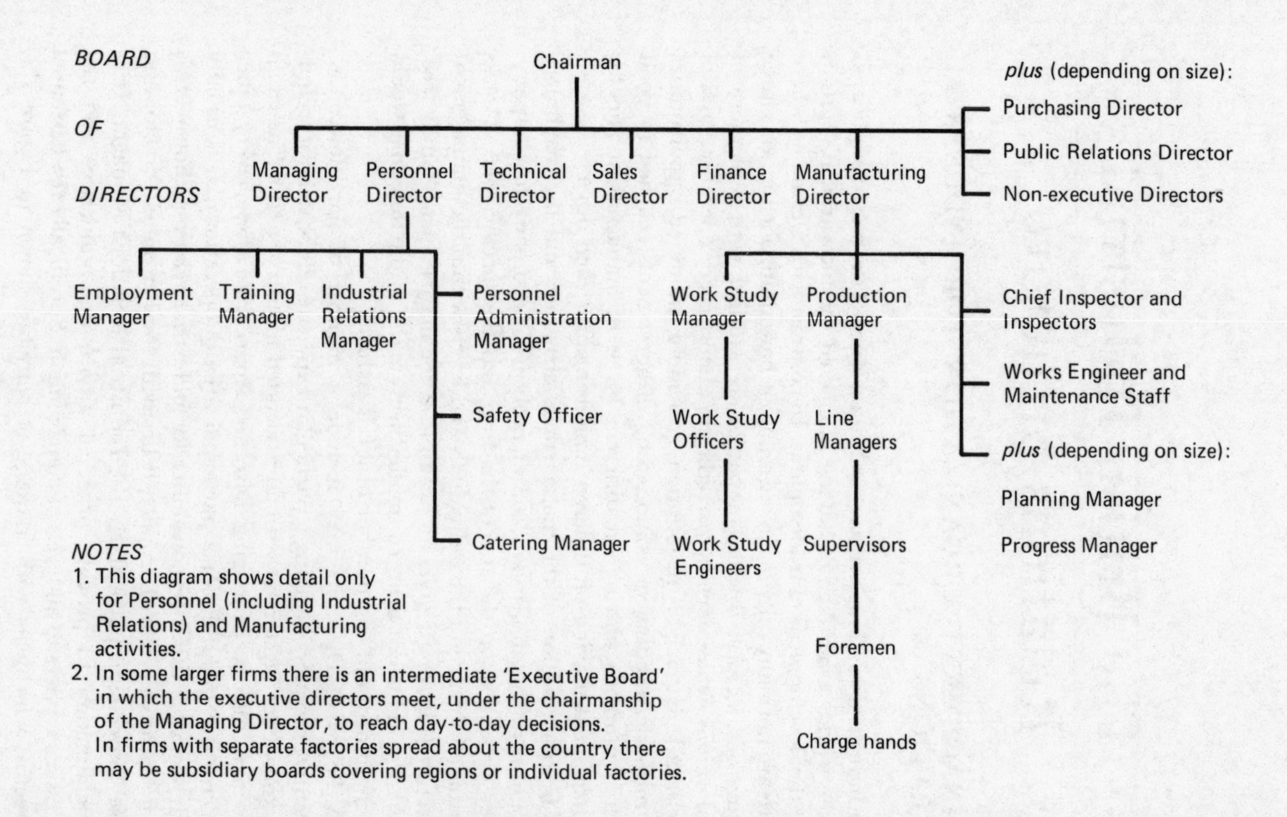

FIGURE 2.1 *Typical management structure of a large manufacturing company*

formulates the industrial relations policy of the firm, in all its aspects. Under him, the Industrial Relations Manager gives advice to the Director on industrial relations strategies, policies and practices, as well as co-ordinating the implementation of the policy. It is also his task to conduct negotiations with shop stewards, works convenors and the trade unions. In practice, however, this formal procedure is sometimes altered, and on occasions may be ignored altogether. For example, the Managing Director can become directly involved – when, for instance, he addresses a works meeting or negotiates directly with the trade unions. The Manufacturing Director is also likely to be involved because he and his Work Study Department are responsible for the scheduling of manufacture and the negotiating of rates for incentive bonuses or piecework. These are usually at the core of day-to-day negotiations between management and the shop floor, and it is when these negotiations fail to reach agreement that they evolve into a dispute. The Manufacturing Director is therefore usually present at meetings between higher management and senior shop stewards.

Informal or 'non-structural' factors can have an important effect upon industrial relations. For example, there may be a clash of personalities between an individual manager and a shop steward, causing the shop steward to try to bypass him altogether. There may be similar difficulties within management or within the body of shop stewards. A particular director or manager may, because of seniority or of real or perceived 'clout' or simply of popularity, carry more weight in negotiations than the others. Such influences can be of great practical significance.

All the levels within the management family tree are involved in industrial relations. Figure 2.1 shows a wide tier of senior managers whose functions are mainly self-explanatory. Some will be constantly involved in negotiations, particularly the Work Study Manager who is in charge of production scheduling; and most get involved from time to time. It is not uncommon in British firms for managers to spend up to a third or half their time on industrial relations, always in the hope of avoiding the expense of negotiations deteriorating into a dispute.

The line managers are immediately under the control of this layer of senior managers, each being responsible for a more specific function, such as the manufacture of a particular product or group of products. The line managers decide upon the composition of the workforce, gangs or teams and distribute the work between them. The supervisors and foremen are the front line junior managers. In a well-run firm, an enormous proportion of the problems over piecework or bonus rates or production schedules are resolved by negotiations on the working level between shop stewards and line or junior management.

EVOLUTION OF THE TRADE UNIONS

Historically, most British trade unions have evolved consciously in one of three categories: craft, industrial and general unions. Craft unions were those whose members had a particular set of skills, regardless of which industry employed them. Industrial unions on the other hand aimed to recruit all workers in a particular industry, regardless of their skills, or lack of them. General unions were those which recruited their membership regardless of either occupational or industrial boundaries, enrolling anyone and everyone, whatever their skills and industries.

These historical distinctions are still reflected to some extent today and, more importantly, some of their traditional attitudes persist (for example to protect their status and independence as do the train drivers in ASLEF) but in most unions the distinctions have become blurred, so alternative all embracing terms such as 'occupational' and 'white-collar' or 'non-manual' have come into use. Occupational unions are less uniform than the craft unions confined to a set of skills. Non-manual and white collar are more easily recognized. Others classify unions as 'vertical' or 'horizontal' – a vertical union obtaining its membership from workers with a common industrial background, regardless of their skill or occupation, while a horizontal union recruits its members from within a particular grade or grades of workers, regardless of the industry in which they work.[1] Yet others have described unions as being either 'open' or 'closed' – a union being 'open' if it is actively recruiting members from, or seeking mergers with other unions in, new occupational or industrial areas; and 'closed' when it concentrates on its existing 'sphere of influence', aiming to make it a stronghold and hence create a strong position from which to bargain.[2]

In practice none of these classifications are fully satisfactory – they exist only as tendencies and are not all-embracing – and only serve to highlight the extraordinary complexities and rivalries of the British trade union movement, in contrast to the relative simplicity of the German system, created from nothing in 1946–47 (with British advice – see Chapter 4). This heritage is one of the handicaps which bedevil the working of industrial relations in British industry.

Although unions existed in embryonic form at least as long ago as the sixteenth century, modern British trade unionism is essentially the creation of the Industrial Revolution when, as one writer remarked, the factory made them possible and the conditions in it made them necessary.[3] These early unions had three main features: they were craft unions; even when they were organized on a national level, they were highly decentralized; and, unlike many continental trade unions, they were not, on the whole, formed for political ends, being content to deal with levels of pay and conditions of work.

Towards the end of the 19th century these three characteristics began to alter. With the development and growth of large general unions the exclusiveness of unions began to disappear. The development of a 'national' level of industrial relations (in which, for example, pay and conditions within large sections of the economy were increasingly determined by industry-wide agreements) strengthened the position of the national leaders at the expense of the local levels. Following a series of unfavourable court decisions, the unions took to political lobbying in an attempt to safeguard their position. It was for this reason that the Trades Union Congress created, in 1871, a parliamentary committee. The experience gained led to a wish for more direct political involvement and so in 1899 the TUC conference voted to set up a Labour Representation Committee. This was later renamed the Labour Party and its influence in Parliament proved its significance with the passing of the Trade Disputes Act by the newly elected Liberal Government in 1906.

Two additional important changes took place. First, during the 1920s, the Labour party supplanted the Liberals as the only realistic alternative to the Conservatives as the party of the government; secondly, from the 1940s onwards, shop stewards began to appear in large numbers as the unpaid spokesmen of workers on the shop floor. By the middle of the 1960s, when wage levels for a significant number of manual workers had come to be determined by workplace negotiations, shop stewards had in many cases replaced the official union machinery as the representatives of the rank and file in the place of work. (see Chapters 9 and 10.)

The dual development of the trade union movement along both industrial and political paths has given the movement a somewhat schizophrenic personality. These two types of activity have presented the trade unions with both direct and indirect political power. Their direct political power is concentrated inside the Labour party. Unions affiliated to the party – which includes most major unions – raise a political levy from their members, which is paid into Labour party funds. The 63 affiliated unions provide around 75 per cent of the party's national finance and up to 95 per cent of special general election appeals. These unions also control around 90 per cent of the votes at the party conference, automatically fill 12 of the 29 seats on the party's National Executive Committee, (NEC). They have the majority say in the choice of six others, and hold 40 per cent of the votes in the electoral college which is used for electing both the party leader and deputy leader. The unions also sponsor directly around a third of the Parliamentary Labour Party (PLP) – in May 1979, 134 out of 269 Labour MPs were union sponsored. In 1972 a TUC-Labour party liaison committee was established. This is made up of nine members from the PLP, led by the party leader, ten members from the NEC, and seven from the TUC, led by the General Secretary. It meets once a month and has become the most powerful decision-making body within the Labour movement.

The union, however, may exercise even greater power indirectly, on both Conservative and Labour Governments, by the use of their industrial muscle.

TRADE UNION ORGANIZATION FOR INDUSTRIAL RELATIONS

As with management structures, so the workforce structures for industrial relations differ widely in different firms and industries, depending upon the number and size of unions represented at each level or area within the firm, as well as upon the functions attributed to each of the parts of the structure. Figure 2.2 shows one example of a trade union family tree. In this example there are four unions operating, but the number may be many more or many less; also, they may vary greatly in strength, with one or two usually dominating (often the AUEW and the TGWU) and a plethora of small ones. This will be illustrated in some of the case studies.

The system of trade union representation is usually organized on a departmental or workshop level, with the union membership electing one of their number as a shop steward to speak for them. There are few formal rules covering the ratio of shop stewards to workers and there are wide variations between levels and between unions, but the commonest scale is about one shop steward per 20–30 workers.[4] Although the duties of a shop steward include union administrative work, such as collecting subscriptions and recruiting members, their main function is to represent the workforce in negotiations with the management over pay and working conditions. All the shop stewards are members of the

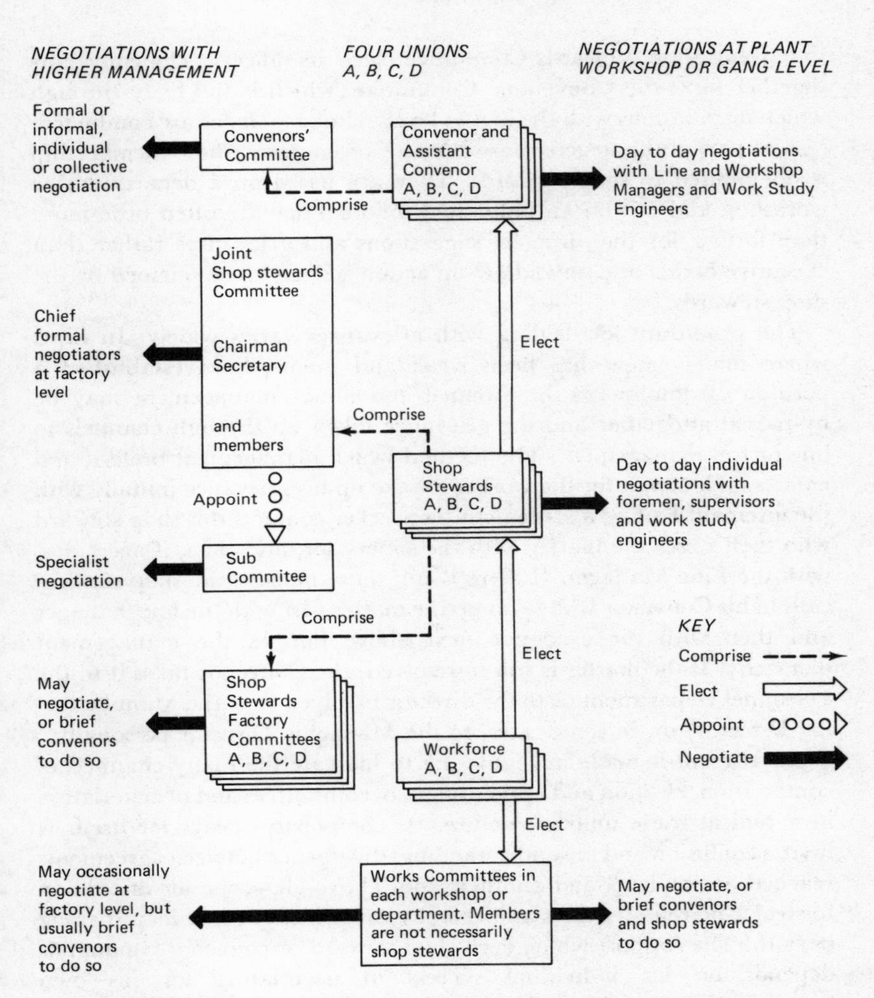

FIGURE 2.2 *Example of a trade union industrial relations structure*

Joint Shop Stewards Committee, while shop stewards within each union form a Union Factory Committee. Within any one factory, each union elects a senior Shop Steward as Convenor. In some unions the Convenor is elected by all the members and in others by the union shop stewards, and it is noticeable that the method of elections is reflected in the attitudes and priorities of the Convenor. If six unions were represented within the firm, there would be six convenors. These convenors are at the apex of shop steward hierarchy, and it is from amongst their number that Joint Shop Stewards Committee elects its officers. The convenors together form the Convenors' Committee, which is the body through which negotiations with the higher levels of management are conducted.

Also part of the structure are Works Committees, whose membership is not limited to shop stewards. They are based on a department or workshop and not on an individual union. They are often little more than forums for the airing of suggestions and grievances rather than executive bodies and any follow-up action is likely to be pursued by the shop stewards.

The procedure for dealing with grievances varies widely. In firms where management has been weak and unions have established a position of dominance the foremen and junior management may be by-passed altogether and the grievance taken up through channels to line or top management. The method which management prefers, and enforces if it can, is for the worker to take up his grievance initially with the foreman. Failing a settlement the worker contacts this shop steward who then raises the matter with the supervisor, and then, if necessary, with the Line Manager. If there is still no settlement the shop steward calls in his Convenor who takes up the matter first with the line manager and then with the executive next above him in the management hierarchy. If the matter is still unresolved, the Convenor takes it to the Personnel Department or to the director involved (e.g. the Manufacturing Director) or, in some cases, to the Managing Director personally.

An attempt is made in Figure 2.2 to indicate the many channels of composition, election and appointment of committees and of negotiation in a typical trade union structure. Its complexity speaks for itself. It invites confusion and misunderstanding; divergence between agreements reached at one level and another; and, above all, scope for one union to steal a march on others and thereby cause bitterness or leapfrogging or both. The support which each shop steward or convenor commands depends on his individual success in negotiating for his own members – though if this is at the expense of too frequent stoppages and loss of earnings it can rebound on him. Add to this the fact that some (mercifully only a few) of them have a wider political purpose in attempting to discredit the firm or the government or system or even to hasten their collapse, and it becomes clear why British managers have to direct so much time and energy to industrial relations, why they so often accept inefficient and restrictive practices to buy peace and why, as a result, the British people, managers, workers and consumers alike, suffer so severely from 'the British disease'.

3 Government Involvement in Industrial Relations

HISTORY

The economic and industrial relations policies of the various political parties and pressure groups have their roots in the history of the British industrial revolution. Until 1824 it was illegal for workers even to combine for collective action against their employers. Then, after half a century of development of trade unions, picketing became legal in 1871 but until 1906 it was largely ineffective because the employer could sue a trade union for any loss or damage he suffered as a result of a strike.

The Trade Disputes Act of 1906 first created what is usually called the 'Golden Formula' under which trade unions and pickets are immune from damages under civil law for their actions provided that the picketing itself is lawfully done. (They are not, and never have been, immune from prosecution for breach of *criminal* law by obstruction, assault, intimidation, etc.) This Golden Formula has remained in essence unchanged since 1906, through a series of acts by Liberal, Conservative and Labour governments. There have been changes in the definition of when picketing itself is lawful (e.g. since 1971 the immunities have been withdrawn from picketing a person's home and from some kinds of secondary picketing) but the principle remains. Such picketing does not itself become a criminal offence but it simply ceases to attract the immunity from civil action. Even when the picketing does fall within the definition of 'lawful', it is only immune if it comprises peacefully imparting or obtaining information or peacefully persuading people to work or not to work, and then only if it is in contemplation or furtherance of a trade dispute.

The first important change came after the failure of the General Strike in 1926 when the trade unions were weak and demoralized and were being deserted by large numbers of disillusioned members. Stanley Baldwin's Conservative Government, catching the tide of public opinion, introduced its Trade Disputes and Trade Union Act of 1927, which outlawed any strike which had any object other than to further a trade dispute in the trade or industry in which the strikers were engaged; and any strike calculated to coerce the government directly or by inflicting hardship upon the community. The Act also altered the conditions under which a union could collect subscriptions for an affiliated political party. It had since 1913 been lawful for a union to levy such subscriptions for a political fund provided that this was kept separate from the main fund to finance industrial action etc and provided that every member had the right to 'contract out' of paying such a levy. Under the 1927 Act it became unlawful to levy subscriptions to the political fund unless the member had specifically 'contracted in' by signing a form expressing the wish to subscribe.

In 1940, under pressure of war, the Government passed the National Arbitration Order outlawing all strikes and lockouts. The Labour Government elected in 1945 restored 'contracting out' for political levies but, to many of its own members' surprise, it kept in force the National Arbitration Order outlawing strikes and lockouts until 1951. Thus, the six years from 1945–51 were the only peacetime years in recent history during which striking has in itself been banned by the law. Since it encompassed the period of restoration of the peacetime economy, after the war, this was in keeping with the public mood as seen by all political parties at the time.

The long period of Conservative government from 1951 to 1964 saw an economic boom. During the decade 1950–59 real wages rose by 40 per cent and by 1960 production was growing at 3 per cent per year. There was a generally cooperative relationship between employers and trade unions which were at that time led by moderate rather than militant Presidents and General Secretaries. This was a time of satisfaction with growing prosperity. Official strikes were rare but there was still much more disruption than in Germany because the cooperative relationship between management and union officials led to some frustration amongst militant shop stewards, who realized that their power and following depended upon their being seen to obtain better wages, bonuses and working conditions for their members – in other words, to fight the management and win. If there were no conflicts they seemed to have no role. They therefore confronted management with demands at shop floor and plant level in excess of nationally agreed wages or conditions and, when their demands were not granted, called unofficial strikes. These, known also as 'wildcat strikes', had become a serious problem by 1964, when the Conservatives failed in their attempt to have a fourth term of office and Labour won the election.

Faced with growing public exasperation with wildcat strikes and with the trade unions in general the new Prime Minister, Harold Wilson, set up a Royal Commission in 1965 to investigate the problem – The Donovan Commission – which made its Report in 1968.[1] Based on this Report the Government in 1969 launched a Draft White Paper for discussion, *In Place of Strife*,[2] setting out various legal restraints on unofficial strikes in default of statutory conciliation and arbitration procedures (as already applied in Germany – see Chapter 4). This, however, blew up in the Cabinet's face, partly because of opposition by Trade Union sponsored MPs in the Cabinet and partly because the Conservative MPs, failing to realize that it was in their party's interest for such legislation to be enacted by a Labour Government, seized what they saw as an opportunity to split the Labour Party.

They paid heavily for their shortsightedness. Under the leadership of Edward Heath they accomplished the defeat of the Wilson Government and passed a very similar Act themselves (the Industrial Relations Act, 1971). Like *In Place of Strife*, this Act aroused the most intense opposition from the Trade Union movement strongly supported by an angry and embittered Labour Opposition in Parliament.

The 1971 Act provided, amongst other things, that

(a) Workers had a statutory right to belong or not to belong to a trade union, thereby in effect making the closed shop inoperable.
(b) Trade union immunities under the civil law became conditional on the unions being registered, this registration in turn conditional on the union satisfying the Government Registrar that its rules contained minimum safeguards for members' rights. Unregistered unions would be liable for unlimited damages for inducing breaches of any contract.
(c) Collective agreements became legally enforceable.
(d) It became illegal to picket a person's home (this is one of the few provisions to survive after the repeal of the Act in 1974 by the Labour government)
(e) The Government introduced a new Industrial Court which had the power, if the Government so requested, to impose a 60-day cooling-off period where industrial action was likely to injure the national economy, imperil national security or create serious public disorder. The Government could also require a strike ballot for similar reasons or where the livelihood of a substantial number of workers was endangered.

These provisions were in accordance with the manifesto on which Heath was elected and with the wishes of a substantial majority of the public as reflected in their electoral vote and in public opinion polls. Nevertheless, the Government learned the bitter lesson that non-cooperation by the trade unions could be an effective counter to industrial relations legislation. They also under-estimated trade union solidarity and overlooked the now well-proven fact that individual trade union members may express a personal view in a public opinion poll or in a secret ballot but the reverse view collectively as union members or under the emotional pressure of a trade union meeting. At its Annual Congress in September 1971 the TUC resolved by a fairly small majority (5½ million to 4½ million card votes) to instruct member unions not to register. By the end of the year the Government was faced with mass non-compliance; apart from a handful of small unions (numbering some 400 000 members) the rest of the trade union movement, with nearly 10 million members, declined to register.

The unions themselves paid a price for this. Unregistered unions lost their eligibility for tax relief on provident funds amounting in the case of the TGWU to about £750 000 per year. The same union was later fined £50 000 for failing to comply with an order to withdraw an unlawful picket.

The Act, however, played into the hands of trade union militants who wanted to clog the legal process by mass disobedience or to arouse passions by creating martyrs. In June 1972 five shop stewards at the London Docks openly defied another Court Order with the intention of being imprisoned, which they duly were. Faced then with a national strike, the Government was saved from a crisis only by the intervention of the Official Solicitor, under an obscure law whereby he entered an appeal on behalf of the 'martyrs' (against their will) for their release.

Later, the government used its power under the Act to order a cooling off period and a secret ballot in the hope of averting a national rail strike but this too failed ignominiously; the railwaymen voted by five to one to support the strike and the Government succumbed and gave them almost all they were asking.

It soon also became clear that employers, who knew that they had to maintain a working relationship with their unions, were reluctant to use their rights under the Act to sue them for damages. More days were lost in strikes in 1972 than in any year since 1926. By the end of 1973 the Act was more or less in abeyance, due to reluctance by the Government and employers to use the power it gave them, and to passive resistance by the unions. In February 1974, challenged by a national miners strike coinciding with an energy crisis when the Arabs quadrupled the oil price, the Heath Government called a General Election and was defeated.[3]

The new Wilson administration not only restored but generally increased trade unions powers in the Trade Union and Labour Relations Act of 1974 and the Employment Protection Act of 1975. In particular they extended the immunity of picketing to cover commercial contracts as well as employment contracts which meant, in effect, that any kind of secondary picketing or secondary blacking was covered, whether the goods or premises affected were connected with the dispute or not; in other words, miners or teachers or railway workers who picketed a steel works or drivers or civil servants who picketed a factory or a hospital became immune from civil damages provided that they could substantiate a claim that their action was in furtherance of their dispute.

The powers given to unions by these laws were demonstrated in a series of transport and public service strikes in January and February 1979. Widespread picketing by truck drivers caused 235 000 other workers to be laid off and, in the most severe winter for a generation, striking public service workers left icy roads ungritted and on occasion prevented treatment of patients in hospitals. For the second time in five years in what became known as 'the Winter of Discontent', the trade unions brought down a government – this time of their own Labour Party.

The 1974–79 Labour Government had, however, introduced what proved to be one of the more successful ventures in the industrial relations field – the Advisory, Conciliation and Arbitration Service (ACAS). This was established under the Employment Protection Act, 1975. Of its nine-member Council, three were nominated by the TUC, three by the CBI, and three by the Government in consultation with them. It offered advisory and information services to employers, workers and their representatives; it also provided a means whereby employers and trade unions could be helped to reach mutually acceptable settlements by a neutral and independent third party. Its essential characteristics were that its use was voluntary and that agreeements reached in conciliation were the responsibility of the disputing parties. Arbitration was also voluntary. An arbitrator or board of arbitration examined the case for each side and made an award. Such awards were not legally binding. In a pay dispute ACAS could also provide mediation services if asked to do so by those party to the dispute. It acted as an independent third party in such mediation and made recommendations for a settlement or suggested a basis for further discussions.

Inevitably, successful conciliation is not news whereas failure, and the strikes which follow it, get maximum publicity. Within industry and the unions, however, ACAS acquired trust and confidence. Should Britain ever follow Germany and the USA in making a specific process of conciliation a precondition for legal recognition of immunity for unions and strikers in a dispute, ACAS could provide a body with the general acceptance and respect to carry out such a process.

MRS THATCHER – 1979

In the wake of 'The Winter of Discontent' a Conservative Government under Mrs Thatcher was returned in May 1979 with a mandate for radical reform of industrial relations. Analysis of the vote revealed that a large part of her vote was from manual workers and that the trade unions had become intensely unpopular even amongst their own members. In a *Times*/ORC public opinion poll published on 21 January 1980, 83 per cent of the sample considered that the unions had too much power and 86 per cent that pickets should only qualify for immunities when picketing their own place of work; 73 per cent considered that trade union leaders should be elected by secret ballot and 85 per cent that there should be secret ballots before strikes.

The poll also noted which of the sample were trade union members, and of these, 79 per cent considered that picketing should be confined to their own place of work, 76 per cent that union leaders should be elected by secret ballot and no less than 87 per cent that they should not be called out on strike without the opportunity to vote whether or not they wished to do so in a secret ballot.

Mrs Thatcher's economic philosophy was to reduce taxation, to cut public expenditure and to attack inflation by restricting the money supply, thereby forcing industrial firms to become more competitive by lowering unit production costs.

The results were a large increase in bankruptcies amongst small firms, unable to borrow enough at the high interest rates to maintain their cash flow, and high unemployment. By the time she sought re-election in the General Election of June 1983 unemployment had reached 14 per cent but the average earnings of the other 86 per cent in work were the highest ever in real terms. Her strategy was presumably based on a gamble that, in a secret ballot, enough of these 86 per cent who had 'never had it so good' would vote to re-elect her for a second term, notwithstanding their joining publicly in the almost universal outrage over high unemployment. She may also have judged that the more this outrage was expressed in public service strikes the more unpopular the unions would become. The gamble paid off in her 1983 landslide victory.

Nevertheless, Mrs Thatcher was careful both in 1979 and 1983 to proceed cautiously and not to bring about a confrontation with the trade unions too early in her term of office. In 1979 her first Employment Secretary was a known conciliator, James Prior. His 1980 Employment Act was unexpectedly moderate (and criticized as such by the 'hard right'). Nevertheless, it did restrict the legal immunity for pickets only to those picketing their own place of work or the premises of the 'first supplier or first customer' of the firm in dispute – the direct supplier of materials, components etc or the direct recipient of the firm's products. Even this – the only form of secondary picketing to qualify – was only immune if the action did not interfere with the *commercial* contracts of firms not in dispute (as distinct from the employment contracts of their employees).

Grounds for exemption from dismissal on account of non membership of a closed shop union, previously confined to religious or conscientious objection, were widened by the 1980 Act to include any deeply held personal conviction. New closed shop agreements would require an 80 per cent vote of approval by union members concerned; and workers who were not already members of the union when the closed shop was formed would be under no compulsion to join it. Workers were given right of appeal against action arising from closed shop agreements to a tribunal which could award compensation. This right of appeal extended to appeals against dismissal on account of direct or indirect industrial pressure.

The 1980 Act also made public funds available for secret ballots for election of union officials, for amending union rules or for calling or ending a strike or other industrial action.

After this Act became law, Prior was replaced as Secretary of State for Employment by a harder line ex-trade unionist, Norman Tebbitt, whose 1982 Act went much further. Tebbitt aimed to 'make the closed shop almost a thing of the past'[4] not by making it illegal (as it is in Germany) but by tightening the conditions for its legality. Dismissal for non-membership of a closed shop union would be classed as unfair unless the closed shop had been supported or endorsed in a secret ballot within the past five years by either 80 per cent of those eligible to vote or by 85 per cent of those who actually voted. The minimum compensation for unfair dismissal was fixed at £2000 for those not seeking reinstatement or £14 000–17 500 for those who did seek reinstatement, with provision for special awards up to £20 000. The Act also provided for retrospective compensation for anyone dismissed since 1974 for non-membership of a closed shop if that dismissal would have been classed as unfair under the 1982 Act.

If a trade union applied pressure on an employer to dismiss an employee unfairly, the tribunal could award compensation to be paid by the union and the employee could in such cases be a party to the tribunal proceedings. Other kinds of pressure to compel trade union membership also became unlawful; so did 'union-labour-only' contracts or pressure to impose such contracts.

Dismissal of strikers became legal not only if all those on strike were dismissed (as was already the case) but also of any individual who failed to return by a time stated by the employer after not less than four days notice.

State funds were made available for secret ballots on wage offers as well as for purposes covered under the 1980 Act.

Trade disputes were to be classed as legal only if they were between worker and employer (i.e. not if they were between worker and worker or between union and union); even then they were only legal if they were related wholly or mainly to terms or conditions of employment or, in the case of a foreign dispute, if UK employees were likely to be affected by the result of that dispute.

Unions would have no immunity from damages in disputes classed as unlawful and maximum scales of damages were set out in the Act, ranging from £10 000 against a union with less than 5000 members up to £250 000 against one with more than 100 000 members. Damages and costs could (with some exclusions) be recovered by attachment or requisition of property.

To avoid the spectacle of trade union 'martyrs' being imprisoned for refusing to pay fines for breach of the law, the 1980 and 1982 Acts placed the onus on employers to sue for damages where unions or individual strikers had forfeited their immunity. As with the 1971 Industrial Relations Act, many employers were reluctant to sue. In 1982, large numbers of trade unionists defiantly forfeited their immunity by unlawful secondary picketing and by striking in sympathy with health workers during a long strike against the National Health Service, but virtually no one sued. Nevertheless, the power to claim enormous damages in future cases remained like a Sword of Damocles over the unions, ready to be let fall by an employer whenever the time seemed ripe. This was, perhaps, the way the Government preferred it to be. The numbers of days lost by strikes in 1981 and 1982 were amongst the lowest on record, though the main cause of this was probably fear of unemployment rather than the new legislation.

MRS THATCHER – 1983

At all events, Mrs Thatcher's strategy was a political success. With the opposition vote split between the Labour Party and the Liberal/SDP Alliance, Mrs Thatcher was re-elected in June 1983 with a massive overall majority, increasing from 43 in 1979 to 144 in 1983. The percentage of trade union members who voted Labour fell from 53 per cent in 1979 to 39 per cent in 1983, while 32 per cent voted for the Conservatives and 28 per cent for the Alliance (which proposed measures for trade union democracy by secret ballot very similar to those of the Conservatives). The Labour Party got its lowest percentage of the national vote for more than half a century (28 per cent).[5]

Interpreting this result as a mandate,[6] Margaret Thatcher and Norman Tebbitt lost no time in publishing a White Paper containing *Proposals for Legislation on Democracy in Trade Unions*.[7] As had been the case with the 1980 and 1982 Acts the TUC was invited to discuss these proposals in draft but, whereas in 1980 and 1982 they had refused, in 1983 they agreed to do so. Once again – avoiding the errors of 1971 – it was proposed that enforcements should not be under the criminal law but by extending the right of employers and others aggrieved by 'unlawful' trade union actions to sue for damages under civil law. The 'others' included trade union members themselves, the company's suppliers and the company's customers.

The immunity of unions would, under the new Act, be forfeited unless there had been:

(a) Secret ballots of union members, every 10 years, on whether or not they wished their union to continue with a political fund.
(b) Secret ballots of members to elect the governing bodies of their union every five years (with provisos as below).
(c) Secret ballots of those directly affected before calling an official strike or making an unofficial strike official.

Regarding (a), this was a modification of an earlier proposal to reintroduce the 1927 requirement to ask each member to 'contract in' before raising a political levy from him. Tebbitt's reasons for dropping that were presumably because it would be seen as a deliberate attempt to impoverish the Labour Party, already much poorer than the Conservative Party; also because if union members had to contract in before paying a levy to the Labour Party it would be only reasonable to demand that companies contributing to the Conservative Party should also only be able to do so from a separate political fund to which shareholders would have to contract in. This would be extremely difficult to enforce in view of the enormous number of companies in Britain; it might also deprive the Conservative Party of a large part of its funds.[8]

Regarding (b), members of a trade union's national executive would have to be elected or reelected every five years if they had a vote on that executive. Many general secretaries do not have such a vote in which case they would not need to be elected. Most union presidents, however, do have at least a casting vote so they would be subject to five-yearly elections. Such elections could, with certain safeguards, be conducted at the work place; or by postal ballot, for which government funds could already be claimed under the 1980 Employment Act. If any union refused to hold a ballot or if any President or national executive member declined to submit himself for election when his five years expired, it would be open to any union member to seek an injunction against his remaining in office. If the injunction were granted and he ignored it, he would be liable to action for contempt of court.

While there were strong protests from many union national executive members it was not easy for them to argue that they should enjoy their privileges and immunities under conditions any less democratic than those of MPs, who do have to submit themselves to election by secret ballot at least every five years before they can continue to exercise their power or enjoy Parliamentary immunities.

Regarding (c), it would not be illegal to call a strike without a ballot and the result of the ballot would not be binding on the union but, if anyone sued the union for damages inflicted by an official strike, the union would have no immunity from such damages unless it could show that it had conducted an adequate ballot which gave majority support for the strike. The legislation would exclude unofficial strikes, presumably because of the difficulty of picking the people to sue (individual shop stewards and pickets?) and of the inflammatory effect of martyrdom of individuals, unable or unwilling to pay damages, being sent to prison. Nevertheless, the legislation could lead to an increase in unofficial strikes (as in the 1960s), publicly discouraged but privately supported by the union's national executive and district officials.[9]

Strike ballots imposed or induced by governments have a history of rebounding against them. The only one enforced by the Heath Government went heavily in favour of strike action (see page 22) and, of the

155 ballots imposed by the US governments under the Taft-Hartley Act since 1948, the workers voted for a strike in all but eight of them.[10]

On the other hand, there is every reason to believe (from public opinion polls – e.g. page 24 – and from General Election voting patterns in 1979 and 1983) that the great majority of the public and especially of trade union members themselves do wish to have the right of a secret ballot before their livelihood is interrupted by a strike.

Although the legislation proposed in the July 1983 White Paper did not differentiate between public service strikes and others, the Government gave an indication of its intentions on 27 July when the Prime Minister announced that their would be an independent review body for nurses and other professional medical workers but that any groups which resorted to industrial action would be excluded from the scope of the review body's recommendations. The implications of this proposal will be discussed more fully in the context of the public services as a whole (Chapter 11).

At some later date in the 1983–8 Parliament the Government is expected also to withdraw trade union immunity from strikes which are in breach of procedural agreements.

In November 1983, the 1980–82 legislation was seriously tested for the first time. Arising from a closed shop dispute in the Stockport print works of the *Stockport Messenger*, the National Graphical Association (NGA) put secondary pickets on another of the company's print works at Warrington. The proprietor, Eddie Shah, obtained an injunction but the picketing continued. The Court fined the NGA £50 000 for contempt and, when they refused to pay, imposed further fines and ordered the seizure of all their assets (£11 m). The NGA thereupon halted the national newspapers who also sued. Mass pickets of 4000 at Warrington became violent and injured 25 police. The TUC General Council, conscious of the adverse public reaction, refused by 31 votes to 20 to support them as they were clearly breaking the law. The NGA gave way, paying £675 000 in fines to purge their contempt.[11] No one went to prison and there were no martyrs. Thus far, Mrs Thatcher's laws seemed to be working.

Legislation to control trade unions and strike action, however, is unlikely to succeed in the long run unless a reasonable degree of goodwill is established by greater involvement of the workforce in the running of their companies or services as a *quid pro quo* and in this respect Britain lags far behind West Germany.

4 Germany

ORIGINS OF SUCCESS

Worker participation is nothing new in Germany. Works Councils date from the Weimar Republic in the 1920s, though both they and the trade unions were suppressed by Hitler. In 1945, the shock of total defeat and massive destruction of industry drew out in the Germans a determination, both individually and nationally, to survive and recover.

This was stimulated in two respects by the Marshall Plan in 1947. First, the Germans were given the capital with which to rebuild. Secondly, they had to do so under supervision of the Allied Control Commission (USA, USSR, Britain and France) which governed Germany until 1949 and was charged with ensuring that they did not revive their armament industries. Faced both by the enormity of the task and the desire to maintain their pride in the face of their conquerors, German management and labour closed ranks.

Initially a small number of Germans attempted to form radical trade unions but these attracted little support and quickly withered. Cooperation and reconstruction were the order of the day. Moreover, they had lived for 12 years under Hitler's National Socialism and had seen both the brutality and the shortcomings of the Soviet system during their partial occupation and as demonstrated by the ponderous inefficiency of the Red Army despite its vast numbers. They looked for salvation to democracy and enterprise. This extended to the main stream of trade unions.

> Whoever knows the unions in their everyday work cannot doubt that they are a firm component of the modern organized capitalist system. They increase its flexibility and stability considerably.[1]

Ironically it was the British trade union movement under a Labour Government which advised the Germans on creating a trade union movement free from the seeds of the British disease – just as it was a British delegation which advised them on designing their Parliamentary electoral system which has proved so much more stable and effective than our own. On these two foundations were built the German economic miracle which left the British far behind.

One of the British trade union delegation remained active in British public life until the late 1970s, Mr Robert Edwards, MP, then General Secretary of the Chemical Workers Union. Most of the Control Commission held military ranks so the delegation was asked to appear in the guise of colonels, majors and captains. Edwards insisted on going as a civilian and a trade unionist and quickly got on terms with his German opposite numbers. He, along with his fellow trade unionist Ernest Bevin, then British Foreign Secretary, knew very well what kind of things to encourage and what to avoid.[2]

THE GERMAN TRADE UNION MOVEMENT

The structure of the German trade union movement took final shape in 1949, in time for the foundation of the Federal Republic of Germany. It was headed by the Deutscher Gewerksheftbund (DGB) which, like the TUC, has only a coordinating role. Unlike the TUC, however, the DGB owns a wide range of property and other assets including Germany's biggest insurance company, its fourth largest bank, a chain of supermarkets and the biggest property development company in Europe. The DGB has no political affiliation and, though it subscribes to the SPD, it has, unlike the British trade union movement, no part in party policy or administration.

The individual trade unions are wholly autonomous. Their greatest strength lies in the avoidance of overlap between them. There are only 17 industrial trade unions which now contain 37 per cent of the labour force. All the manual workers in a single industry are in the same trade union, whether they are metalworkers or carpenters, concreters or floor-sweepers. The white collar workers are members of another union – again only one. Thus, in collective bargaining, be it at regional or corporate level, there are only one or at most two bargainers on the labour side. Collective bargaining agreements are only valid if reached between employers and a recognized trade union, and are enforceable by law.

There is no equivalent of the Joint Shop Stewards Committees found in Britain. Officially there is no trade union representation as such at plant level[3] but in practice 86 per cent of Works Council members do also hold a trade union office. The Works Council can also invite permanent trade union officials from regional or other trade union headquarters to attend their meetings if 25 per cent of council members vote to do so. Apart from this, however, permanent union officials visit the plant mainly to recruit members, collect dues and promulgate trade union information. Whether or not some of the Council members are trade unionists it is the elected Works Councils, not the trade unions, which conduct negotiations at plant level. These negotiations do include plant bargaining on wages,[4] but are mainly concerned with such matters as rates of work, shifts and other conditions of work as described later in this chapter.

GERMAN INDUSTRIAL RELATIONS LAW

German industrial relations are supervised by Labour Courts, which enforce collective agreements – it is in fact illegal, not just to break a collective agreement but also to induce someone else to break it.

Closed shops are illegal, since any worker enjoys the right either to join or not to join a trade union and cannot be denied work for either doing so or not doing so.

Strikes and lockouts are thus unlawful if they are in breach of a collective agreement; also if they are for any purpose other than the improvement of terms and conditions of employment; so political strikes or sympathy strikes are unlawful. While it might appear easy to claim that such a strike is concerned with terms and conditions of employment it may not be so easy to find credible demands for improvements which do not fall outside a collective agreement currently in force, and in practice political motivation of strikes is very rare.

German law differentiates between disputes over rights and over interests. Strikes are legal only if over interests – e.g. over a deadlock in collective bargaining. They are unlawful if concerned with rights – i.e. the interpretation of existing legal rules – which must be resolved by the Labour Courts. Since, moreover, it is illegal to strike over any matter currently covered by a collective agreement, it is unlawful to strike in order to force *future* wage changes until the current collective agreement has run out.

There is no specific law to ban secondary picketing but it would almost certainly fall foul of one of the other laws mentioned above (e.g. it would not be concerned with the terms and conditions of employment of the strikers) so in practice it does not occur.

There is provision in Germany for compulsory conciliation procedures.

These normally take at least six weeks, and it is unlawful to strike until these procedures have been exhausted.

All the DGB unions except one require a secret ballot of workers involved and in most cases they can only be called out on strike if the vote in favour is at least 75 per cent. This safeguard is popular with German workers.

A strike by workers in a public utility is only legal if the union concerned gives notice of the strike and sets out the measures necessary to maintain essential services. Full time established public servants do not have the right to strike.

Heavy damages may be awarded against a union for an illegal strike. Any individual who strikes unofficially may also be sued for damages, and unofficial strike action is a valid ground for dismissal.

WORKS COUNCILS

The low level of strikes in Germany and the acceptance of what British trade unionists would regard as restrictive industrial relations laws derive largely from an effective system of worker participation which is enforced by law. This is based on the Works Constitution Acts of 1952 and 1972.

Any establishment employing more than five workers is required to set up a Works Council with certain defined powers at plant level. These Councils are elected by secret ballot of all employees, both manual and white collar staff, but excluding senior management and executives. Anyone over 18 who has been with the company for six months or more is eligible to stand for election, and candidates are elected for a three year term with provision for reelection.

The size of the Works Council depends on the number of employees in the plant – e.g. 7 members for 151–300 employees and 9 members for 301–600 employees. Plants with more than 600 employees should form Works Committees headed by the Chairman and Vice-Chairman of the Works Council with such other members as they require to deal with the day-to-day business of the Council.

When a company has more than one plant, a Central Works Council is formed, which includes two members (one staff, one manual) to deal with matters affecting the company as a whole but it has no power to overrule decisions by the plant Works Councils.

Works councils are entitled to meet in working hours and the employer is required to provide them with accommodation and cover their costs including secretarial support. He also has to cover their legal costs if a question is taken to the Labour Court or if the Council requires legal advice. Council Members have the right of up to four weeks paid release per year for education and training for their task. Every three months the company is required to give all employees time within working hours for a general meeting with the Works Council to which the employer has to be invited.

The functions of works councils include the codetermination with management of the conditions of work, including working hours, shifts, breaks, holidays etc; piecework, bonus rates, principles of remuneration and systems of payment; measures to prevent accidents, occupational diseases and unemployment; and the introduction of new technical equipment and production methods. The employer is obliged to inform the Works Council of any proposed modification (e.g. reduction of operations, or closure of plant), and the Works Council can demand compensation for any of the work force who will suffer from it. If codetermination does not work (i.e. if the Works Council and the Company cannot reach an agreement) the points at issue must be referred to a Conciliation Committee or to the Labour Court.

Works Councils in the above respects broadly perform the functions which British convenors and shop stewards perform at plant level, though with a more limited wage negotiation role. Minimum wage rates are governed by the collective bargaining agreements reached between trade unions and employers' associations at national or regional levels, and Works Councils are bound by law to observe these, though there is often provision for them to negotiate increments at plant level. Works Councils can also police the collective agreements and, if they consider that the employer is breaking them, they can refer the matter to a Labour Court. If a Works Council were to call a stoppage of work it would, as explained above, be classed as unofficial and all concerned could legally be sued for damages or dismissed. The same would apply to any other complaint that the law was being broken – e.g. safety or health regulations – which would be for settlement by the Labour Courts.

In firms with more than 1000 employees, Works Councils must form an Economic Committee, which meets monthly to examine the financial, production and investment situation, consulting with management and keeping the Works Council informed. This Committee has not more than seven members of whom at least one must be a member of the Works Council. The other members are chosen for their expertise in the appropriate fields and may often be members of the company's management.

Works Councils also have codetermination rights over appointments, gradings, regradings and transfers of those they represent and they have to be informed in advance of any lay-off, dismissal or redundancy whether voluntary or otherwise. If this is not done the lay-off or dismissal is void until the decision has been referred to a Labour Court. In practice, most firms have printed forms on which they give regular information to Works Council of whom they intend to regrade, transfer, dismiss, etc., with explanations where appropriate. The Works Council have one week in which to discuss these, after which they are referred to the Labour Court. This system normally works quite smoothly.

THE SUPERVISORY BOARD

Two tier boards have been mandatory in German public companies since the Companies Act of 1884. In the 19th Century the shareholders elected a Supervisory Board which appointed a Management Board. The Co-determination Act of 1951 required coal, iron and steel companies with more than 1000 employees to set up parity representation on the Supervisory Board, whose minimum size was 11. Of these, 5 represented the work force, 5 the shareholders and the 11th was neutral, agreed by both sides, with a casting vote. The Works Constitution Act of 1952 made similar regulations for other public companies with more than 500 employees, requiring one third workers' representation on the Supervisory Board and the setting up of Works Councils.

Under the Co-determination Act of 1976 there have been further moves towards parity representation. The size of the Board for companies with 2000–10 000 is 12, for those with 10 000–20 000 is 16 and for those with over 20 000 is 20.

Where there are 12 members, 6 represent the shareholders and are elected at the annual shareholders' meeting; the other 6 represent the work force, of whom 2 are appointed by the unions (manual and white collar) and 4 by direct election by secret ballot. Of the latter, two are elected by the manual workers, one by technicians and salaried staff, and one by junior managerial staff.

The Chairman is elected by a majority vote of the whole Board but, if this is inconclusive, the shareholders elect the Chairman on a second ballot. In Board proceedings thereafter, if there is a tied vote, a second ballot is held in which the Chairman has a casting vote. Thus in cases where the parity principle results in the Board being unable to reach a majority vote, the shareholders' representatives have the ultimate power of decision on a second ballot.

The Supervisory Board elects the executive Management Board. There is no fixed composition of this, but it will include a President (i.e. Managing Director) and Directors for such matters as Finance, Technical, Production, Marketing and Labour. All are subject to election as individuals by the Supervisory Board which also approves their salaries.

CONVERSATION WITH A WORKER-DIRECTOR

Walter Thiemann is one of the two directly elected manual workers' representatives on the Supervisory Board of Westfalia Separator, a manufacturing firm employing about 2000 in Oelde and 1000 in a number of smaller plants. The author spent a morning with him in 1979.

'What do you manufacture?'

'Separators, decanters, clarifiers, milking machines, milk coolers, butter churns – mostly for the wine and dairying trades.'

'Do you export much?'

'Yes, worldwide – from Latin America to New Zealand – and we have factories here in Germany and in France and Brazil. We have one major competitor, a Swedish firm but, as you see from the charts on the wall, we are doing well for sales because we keep our costs down. Thirty per cent of the costs are for labour.'

'Have you had many strikes?'

'There has never been a strike, and the firm has been going for 85 years. We would need a 75 per cent vote to have a strike. Only twice in my time have we had a majority vote for a strike, but neither time was it as high as 75 per cent.

'In those two cases, what would have happened if the workers had gone on strike all the same?'

'Only the union's central executive could override its own 75 per cent rule and call a strike but this has never happened here. Otherwise, men who stopped work would not legally be on strike. They would get no money from either the company or the unions and they could be sacked.'

'I understand that you have just completed a previous term of office and have been re-elected.'

'Yes, with a majority of 78 per cent. I represent the manual workers in the Oelde factory and there is a second manual worker elected by the other plants in Germany.'

'Why do you think they re-elected you?'

'Presumably because they are satisfied. The number of jobs has increased by 50 per cent since I joined the firm, because our prices are competitive, and our wages are higher than the average. We're the biggest firm in Oelde and 10 000 other people depend on us. There is no unemployment in this town and it is my job to keep this so.'

Westfalia Separator have a sales and servicing unit in Britain but no factory. They tried production in Britain but, said the President, it never seemed to make a profit. Though British wages were only two thirds as much as in Germany, production was less than two thirds as much and stoppages per thousand employees, nine times as much.[5] To quote Walter Thiemann again:

> Our trade unions have a very different approach from yours. There is no confrontation. We want high production just as much as management does because the more we sell the more work there is in Oelde.

So we buy their separators and milking machines but the Germans, the French and the Brazilians get the work.

5 Sweden, France, USA and the EEC

SWEDEN

The post-war Swedish success story rivals that of Germany and is based upon a highly centralized Federation of Labour (LO), founded in 1898, and now encompassing over 90 per cent of the nation's industrial work force. There is little direct government interference in industrial relations because both the LO and the Swedish Employers Association (SAF) have shown a joint and consistent determination, ever since the 1930s, to solve their own problems. They established the framework of their relationship in 1938 in their Saltsjobaden agreement which

> established negotiation procedures including grievance procedure; rules which must be observed in case of dismissals and lay-offs; limitations on strikes, lockouts and similar direct action; and a special procedure applying to conflicts jeopardising vital interests of the community.[1]

Since 1952 LO/SAF negotiations have been national in scope. They determine recommendations on wages, working hours and fringe benefits for periods of three years. Their agreements form the basis for contract negotiations between individual unions and companies, not only for those affiliates to LO and SAF but also to others. The Government's role is restricted to one of mediation when extended disagreements occur.

These industrial relations processes are founded on long established laws governing mediation in labour conflicts (1906 and 1920), collective agreements and the Labour Courts (1928) and freedom of association and negotiation (1936).

The Labour Court (established by the Collective Agreements Act of 1928) originally comprised two members each from the LO and the SAF and three appointed by the Government. In 1947, reflecting the growth of the white collar element in industry, a representative of the Central Organisation of Salaried Employees (TCO) was added and, since 1966, a government representative replaces one of the SAF delegates when the dispute involves the state as an employer.

The unions in Sweden have always prided themselves on their willingness to accept a high level of responsibility for the national economy and this was reflected in the remarks of Arne Geijer when he was Chairman of the LO:

> It is no trick for us to stand society on its head. It is easy, in a ticklish conflict situation, to act so that society cannot function. But such conflict is no longer a conflict against employers but rather against society and ourselves. At present, a serious labour market conflict affects all citizens and our members to a greater extent than employers. LO alone now organizes every fifth Swede, between half and one third of all persons of working age. We are no longer a small social minority. Society is to a large extent ourselves and our families . . . I believe that, however we act in the future we must, willingly or not, take the national economy into consideration.[2]

It is chastening to compare this attitude with Mr Arthur Scargill's expressed wish to 'paralyse the nation's economy' in 1972.[3]

The result of this attitude of the LO has been a remarkable record of industrial peace in Sweden, rivalling even that of West Germany. (see Table 5.1)

TABLE 5.1 *Working days lost per 1000 employees*[4]

Year	*Sweden*	*Germany*	*UK*
1972	3	3	1,081
1974	16	49	647
1976	7	26	146[a]
1979	7	22	1,291
1980	1,148[b]	6[b]	531[b]
(AVERAGE)			
1971–80	163	54	575

NOTES

[a] Lowest figure from UK from 1971–80.

[b] All figures for this year provisional.

FRANCE

In Britain, the trade unions appeared before socialism and the British Labour Party was always (and remains) dominated by the trade union movement. In France socialism appeared before the trade unions. This goes a long way to explain the difference between the trade unions in the two countries.

Trade unions at national level were legalized in France in 1884 but French trade union leaders, unlike British ones, showed little faith in peaceful reform. In its Charter of Amiens (adopted in 1906) the French General Confederation of Labour (CGT) affirmed that:

> The class struggle places the workers in revolt against all forms of capitalist exploitation and oppression, material and moral... (The Trade Union movement) prepares for the complete emancipation which can be achieved only by expropriating the capitalist class.[5]

This attitude has never, however, been widely reflected amongst the French labour force, and has also ensured that both governments and employers have stoutly resisted any extension of union power. This resistance was facilitated by France's late and incomplete industrialization and the continuing predominance of agriculture in French society. The French people have also always displayed sharper political, ideological and religious disagreement than most and the trade union movement has been hampered by internal conflicts between syndicalists, communists, socialists, reformists and Catholics. The movement is politically divided and weak, encompassing only about 20 per cent of the industrial labour force and 10 per cent in private industry. As a result, the individual workers or groups of workers tend to have a more subservient relationship with their employers than in Britain, Sweden or Germany.

The French Constitution guarantees both the right to strike and the right to work; also the right to belong or not to belong to a trade union. Hence the closed shop is illegal though in some sectors such as the national press and the docks there is a *de facto* closed shop.

The right to strike has been limited by the courts' interpretation of the Constitution. Working to rule, restricting output and 'rotational' strikes (ie one group stopping after another to cause a continuous stoppage) are all illegal. So is any action which damages an employer's property or is in breach of a collective agreement.

Most collective agreements are concluded nationally, usually on an industry basis. Such agreements are legally enforceable as civil contracts and are binding on all employers whether union members or not. So are individual contracts between employers and employees.

Strike action in essential services is restricted by law. The police, the armed forces, air traffic controllers and certain other public servants are prohibited from striking. Nuclear power station workers may not take any action which endangers safety. Striking radio and television workers are obliged to maintain a minimum service and striking teachers must keep children under supervision at school during normal school hours. These laws have, however, been broken with impunity by both air traffic controllers and teachers. Industrial action is only lawful if it is in promotion of 'employment interests' so a political strike is technically illegal but, again, the Courts have placed a wide interpretation on 'employment interests'.

Picketing for peaceful persuasion is lawful but it is a criminal offence to impede a worker from exercising his constitutional right to work. There is no specific legislation about secondary picketing which is in practice rare in France.

French law regards the individual, not the union, as responsible for the consequences of industrial action, so there are no specific union rights, obligations or immunities in law. Nevertheless, there have been a growing number of civil actions against trade unions.

There is no legal requirement for pre-strike ballots. Trade union rules – confirmed by ministerial decree – do, however, provide for ballots for the election of trade union officials.[6]

Both the Swedish and French societies have been very successful, though their trade union attitudes and powers show an extreme contrast. This suggests two lessons for British trade unions: if unions are too politicized (as in France) they attract fewer members and become weak and divided, resulting in a successful economy at the expense of a subservient work force – but also at the cost of contributing to the major political upheavals which have plagued French society. If unions are dedicated to cooperation, productivity and responsibility to the community they can not only attract massive membership and exert a powerful influence on industry and government, as in Sweden; they can also at the same time achieve prosperity for their workers and a tranquil society. British trade unions fall between the two, and there is in consequence greater disruption and less prosperity.

THE USA

Industrial relations in the USA again show a marked contrast with others. Though stoppages are as high or higher than in Britain, the huge underlying industrial strength arising from her national cohesion, with ample raw materials and transport and a common market, have enabled her to carry this disruption and still remain a prosperous society, though a very much more violent one than the British, in every field including industrial relations.[7]

Only 22 per cent of the US work force belong to trade unions, this figure having fallen from 35 per cent in 1945. Unions are, however strong in engineering, motor manufacturing, mining, railways and steel; also in the public services where union membership is expanding. The US equivalent of the TUC – the AFL-CIO – is rich and influential, but several major unions are not affiliated to it. Strong employers confederations have also kept trade union power in check.

American trade unions, like those in Germany, work in support of rather than against the capitalist system. They are not affiliated to any politicial party but they generally support the Democratic Party, and often help to raise funds for Democratic candidates (though sometimes also for Republican candidates). The labour movement has been riven by internal dissession and personal rivalries (though political ideology is conspicuously lacking) and some sections have been closely associated with racketeering and criminal activities.

The US constitution guarantees certain positive rights (including the right to work but with no absolute right to strike). These rights have been developed by state and federal laws.

The legal framework for industrial relations is embodied in the Labour Management Relations Act of 1947 (the Taft Hartley Act). The Act is administered by the National Labour Relations Board (NLRB) which acts as an industrial court, dealing especially with complaints of 'unfair industrial practice' under the Act.

The Act guarantees the right not to join a union, but does also allow for post-entry closed shop agreements, though dismissal under such agreements is unlawful except for failure to pay union dues. This federal law can be overridden by state law, so in 20 states (mainly in the South) the closed shop is unlawful. In many cases, unions have got round this by setting up 'agency shops' in which employees pay union dues but are not bound by union rules.

Under the Taft Hartley Act, strikes may be either 'protected' or 'unprotected' by the law. Strikes over 'unfair labour practices' by an employer are protected, i.e. strikers may not be dismissed and are entitled to reinstatement at the end of the dispute. On the other hand strikes which employ unlawful means or which themselves constitute an 'unfair labour practice' are unprotected by the law and unions and strikers may be sued for damages. Collective agreements are legally enforceable and generally include no-strike clauses, so strikes in breach of collective agreements are unprotected; so are strikes over disputes other than those between employer and employee, and political strikes (which are in any case very rare as American unions are fundamentally non-political). Strikes by federal public servants are illegal and, though some states permit strikes by some state employees, none permit strikes by police or firemen.

Under the Taft Hartley Act industrial action is only lawful if taken against a party to the dispute, so action against a third party (i.e. secondary picketing) is generally actionable as 'unfair labour practice'. If, however, the third party is carrying out work for an employer involved in the dispute which would otherwise be done by that employer, then the third party loses his neutrality and is viewed by the law as a party to the dispute.

Picketing is protected by the law provided that it is 'informational picketing' and provided that it seeks only to persuade but not to prevent anyone from working. Picketing is unlawful if it obstructs entry to an employer's premises (the courts commonly grant an injunction limiting the number of pickets to two per entrance). It is also unlawful to picket the premises of an employer not in dispute unless he had lost his neutrality as described above. In recognition disputes the union seeking recognition is obliged to apply for a secret ballot to be conducted by the government within 30 days and, if it has failed to do so by that time, picketing becomes unlawful; it is also unlawful if recognition cannot legally be extended (e.g. if employees are already represented by a different union). Where several employers are operating on the same site it is unlawful to interfere with employees of an employer who is not in dispute and, for this reason, it is customary in such cases to establish reserve gates for use exclusively by employees of neutral employers. These gates cannot lawfully be picketed. Picketing which involves 'serious misconduct' (e.g. violence or threatening behaviour) constitutes an 'unfair labour practice' and is thus unprotected by the law, in addition to the behaviour itself constituting a criminal offence.

Federal law does not require pre-strike secret ballots (except in strikes in essential services – see below). Several states, however, provide for such ballots and many unions hold them voluntarily. Federal law does require secret ballots for election of union officials, at least every five years at national level, every four years at intermediate level and every three years at local level.

The best known provision of the Taft Hartley Act concerns strikes in essential services. If the President considers that a strike will affect a whole industry or 'imperil the national health or safety' he may appoint a Board of Inquiry to investigate the facts of the dispute. He may then ask the Attorney General to seek an injunction to restrain the strike and appoint another Board of Inquiry to advise him on the dispute. There is then an obligatory cooling off period of 80 days towards the end of which the NLRB must hold a secret ballot of employees concerned on whether to accept the employer's latest offer. If the employees reject the offer the injunction is lifted. The President then has to submit a report to Congress with recommendations for further action.

In 1981, President Reagan applied this procedure to the Air Traffic Controllers strike and, when the union declined to order a return to work, the union was disbanded and all the air traffic controllers sacked. Alternative air traffic control arrangements were quickly made while new staff were recruited and trained with little disruption and the President defeated the challenge with general public approval.

THE EEC FIFTH DIRECTIVE AND VREDLING

Various attempts have been made in the EEC to encourage better worker participation and communication by member states, notably the Fifth Directive and the Vredling proposals, but little progress has been made.

The Fifth Directive on company law is designed to give the backing of law to worker participation. The idea was first debated in 1972 but bogged down in a series of amendments only one of which was adopted. In May 1982 the European Parliament adopted a report by 158 votes to 109, which proposed to leave discretion to member states to set up either a two-tier system on the German model, or a single tier system with worker directors or a works council. For two-tier systems and for companies with more than 1000 employees, it was proposed that between a third and a half of the members of the supervisory board should be appointed by the employees but there were safeguards to enable employees to opt out altogether if they so wished, and for the ultimate decision-making power to lie with the shareholders – again as in Germany (see Chapter 4). It is, in fact, unlikely that the EEC will be able to impose a uniform system on its members because, amongst other things, some trade unions are nervous about the possible by-passing of the collective bargaining process.[8]

The Vredling proposals originally came out in 1973 but were first debated only in September 1982. There was much opposition and over

300 amendments were tabled. The proposals concern communication in multinational companies operating in Europe. The final draft now proposes that every 6 months a multinational corporation would have to forward information to its subsidiaries giving a clear picture of the activities of the corporation as a whole, including structures and manning, the economic and financial situation, probable developments in production, sales, employment and working methods or anything else likely to have a 'substantial effect' on employees interests.[9]

These proposals will probably be adopted, though with some amendments, since it is accepted that legislation is needed because differences in the rights of shareholders and of employees cause difficulties in establishing multinational subsidiaries, in creating mergers and in the commercial development of the common market. The proposals are in general supported by the Conservative Government in Britain and by most European trade union movements.[10]

6 The Media and Industrial Conflict

THE NATURE OF THE FREE PRESS

The mass media have collectively become the 'joker in the pack' in industrial disputes, and both sides may attempt to harness them to get their point of view over to the public. Because conflict is news, the media, especially television, are very ready to cover it so long as they know where it is happening. In democratic countries the free press will therefore deploy its journalists to seek out any confrontation and they will sometimes wittingly or unwittingly, arouse and exacerbate it.

Freedom to report is sadly something of a rarity. The great majority of the world's population can read and hear and see in their media only what their government thinks fit. In authoritarian countries, whether they are left wing dictatorships, left or right wing one-party states or military regimes, the media are either owned and operated as organs of government, or heavily censored or, more insidiously, controlled by retrospective retribution which the editor knows will come if he incurs the government's displeasure. In free societies the media are often criticized for being commercially motivated, for trying to please capitalist proprietors, and for pandering to the worst taste in order to attract an audience. All of these motivations, however, are healthier than those in a country whose media are controlled by the state, where a journalist's future depends on his ability to persuade his readers and viewers to believe what the government wishes them to believe.

In a pluralist democracy, the conflicting views of sectors and factions of society can only be brought to bear by freedom of speech in a free press. A free investigative press is also essential as the only effective restraint against the growth of corruption and the abuse of power and these, if not restrained, are mortal diseases for a free society.

If journalists are to be free to investigate and disclose they will also be free to act irresponsibly with or without some retribution under criminal or civil law. For employers, trade unionists, government officials or policemen who have to live with a free press in their daily work, it is important that they understand how and why it operates in the way it does.

The prime motivaton of a journalist in a free society, in radio, television or the print media, is to attract an audience. He does this by striking a chord with his regular readers or viewers and, hopefully, with some of those on the fringe of his rival channel or paper so that they switch to his. This is not only a commercial motivation (powerful though that is); it is also a professional motivation. The BBC has no commercial motivation but its editors, producers, presenters and reporters have just the same imperative to attract an audience as their rivals in ITV. A good rating for their programme is their prime source of professional pride; a bad one is not only a humiliation but, if it occurs too often, is also likely to lead to fewer and fewer interesting or challenging assignments.

Striking a chord means saying what the particular audience wants to hear and with which it will agree. This can be unhealthy because much of the British audience has stereotypes of bosses, trade union officials and shop stewards and has prejudices about them. Striking a chord is likely to reinforce these stereotypes and prejudices. The chief sufferers from this are the unions, especially shop stewards and pickets. A pompous official, a hectoring steward or a snarling picket are the stereotypes, so some journalists will hunt for people to interview or photograph who fit these stereotypes. The stereotypes predominate in many news and current affairs reports.

A BIAS AGAINST THE UNION

Most trade unionists are convinced that the media have a bias against the unions. This belief was expressed in two pamphlets published by the TUC in 1979, in the wake of the Winter of Discontent.[1] They quoted a number of what they regarded as slanted and inflammatory headlines during the transport and public service strikes – 'Britain Under Siege' (*Daily Mail*), 'Four Million Children Are Blockaded from School' (*Sun*), 'What Right Have They to Play God with My Life?' (quoted from a heart surgery patient) (*Daily Express*), 'You Name It – They'll Stop It' (*Daily Mail*) and 'Target for Today – Sick Children' (*Daily Mail*).

These headlines have been picked as the most inflammatory, and inflammatory they are. They do, however, reflect what most of the readers of those papers felt at that time. When strikers picketed the children's hospital and took it upon themselves to judge which were emergency cases to be admitted and which were not, most readers were highly indignant. 'Target for Today – Sick Children' precisely struck a chord. The editors know well enough that the public dislike strikes, especially public service strikes, and most of all public service strikes which hurt old people and children. The readers of these headlines therefore probably bought the same paper the next day and reinforced their prejudices.

TELEVISION AND VIOLENT PICKETING

It is an axiom that frustration often leads to aggression – verbal or physical. If pickets feel that the media have a bias against them, and that the true arguments (e.g. about their low pay or the months or years of fruitless negotiation) are not being fairly put before the public, that is the likeliest of all the causes of verbal aggression; and this can develop, as tempers rise, into physical aggression. Thus irresponsible reporting can fuel violence.

Reporting, fair or biased, can create folk heroes. Arthur Scargill was in 1972 an obscure union official handling miners' compensation at the Yorkshire Miners Area Headquarters in Barnsley. By mid-February, television coverage of the violent picketing at the Saltley Coke Depot had made him into a national figure, hated by some but worshipped as a hero by others. He had caught the public imagination and he was news. He has seldom left the headlines ever since.

At Saltley the television cameras put the miners on their mettle. They felt that they had to win, to stop coke coming out of the huge stockpile in the Depot; if they failed other pickets nationwide would be discouraged and the whole national coal strike might fizzle out. The police, on the other hand, had the duty of keeping the public highway to the depot gates open. So a confrontation was inevitable. The television cameras raised the tempo, just as a capacity crowd raises the tempo at Wembley Stadium. Thirty people were injured (16 of them policemen) and the drama made good television.

The camera can both arouse and restrain violence. At Grunwick in North London a very small strike – a dozen pickets outside a factory with a workforce of 400, scarcely noticed for nearly a year – suddenly captured the attention of the media in June 1977. The Socialist Workers Party (SWP) called for a mass picket. London University students, having just finished their exams, flocked to the scene via an underground station 50 yards from the gate. There was a natural marriage between the television cameras and the students, each providing what the other one wanted. The students, and the expectation of a violent confrontation with the police, brought the cameras; the cameras attracted more students and led some of them to 'act up' so that they would grab the news. Between 700 and 2200 student demonstrators appeared every day for ten days and attacked the police cordon, injuring altogether 243 policemen. The pickets themselves – the handful of Grunwick employees on strike – used no violence at all, but the demonstrators claimed to be 'pickets' and were referred to as such by the media. As a result the viewing public unfairly associated the violence with the Grunwick strikers.

The violence actually ended as a result of television coverage of the most violent day of all, 23 June. A demonstrator threw a milk bottle at P.C. Trevor Wilson and it knocked him out. The cameras then focused on the unconscious policeman with a pool of blood spreading from a gash in his neck. These pictures aroused intense public anger and the Grunwick pickets themselves were appalled. They denounced the action and told the student demonstrators to 'get off their patch' as they were ruining their case. The damage, however, was done. A few months later, having lost virtually all public sympathy, they had to abandon the strike with none of their demands met.

The presence of the television cameras originally helped to arouse the violence but also, eventually, to end it. The reporting of it contributed to the growing unpopularity of the trade unions as a whole, which was very rough justice as those using the violence were not the strikers at all.[2]

JOURNALISTS AND THE PUBLIC

The journalist in a free society has the duty to investigate and disclose but he also has the responsibility as a citizen to do so fairly and honestly, and not in such a manner as may inflame violence and put lives at risk. This duty and this responsibility may often conflict.

If industrial conflicts are to be resolved without violence and without undue disruption of the life of the community, responsible reporting of them may be a decisive factor. A good journalist will, in particular, try to cover the background of a dispute so that public judgements will be based fairly on the facts and not on prejudice or emotion. This is, however, more easily said than done. A report on the main television news normally lasts only one or two minutes and very seldom more than five or six, so there is not time for much background. A longer current affairs programme may cover the background but this will be watched by very many fewer people, and not by the same people, e.g. pickets coming off duty after a hard day will see only the main news report and may be incensed by the lack of background of their case and by the disproportionate stress on violence and confrontation. This, however, is inevitable because, if during 12 hours of picketing there have been two minutes of violence it will be these two minutes which will get on the news. That is part of the price of having a free press, but the price of not having one would be greater still.

During industrial disputes, trade union representatives seem to be more ready to be interviewed than employers.[3] When the two sides are invited to confront each other in front of the camera, this may provide good television but is likely to exacerbate rather than resolve the conflict, since neither side will wish to be seen publicly giving ground so attitudes may be hardened and positions taken up from which it will be hard to retreat in proper negotiations later. Constructive negotiation cannot be done in public.

On balance, because industrial conflict is news, the media probably do more to exacerbate it than to resolve it. In the long run, however, a better informed public will act more sensibly and responsibly, so the onus lies on the journalists to act responsibly, in awareness of the consequences of their reporting, and on employers and trade unionists to be prepared to speak to the media and to use their opportunities constructively. If they choose to use the media to drive a harder bargain or to arouse passions, the conflict will get worse. If, however, both sides wish to resolve the conflict, wise use of the media can help them to do so.

Part II
The Price We Pay

7 Costing Industrial Disputes

THE PROBLEM

Assessing the cost of an industrial dispute is at best an imprecise operation and is all too often done with calculated dishonesty as material for propaganda, to arouse passions, to alarm or to discredit. For these purposes costs are sometimes exaggerated by all parties in the dispute in their different ways. Alternatively the costs have on occasions been deliberately underplayed in advance of a strike by, for example, a convenor who knows that if his members knew what it was going to cost them they would never follow his call to strike.

In the days when the strike weapon was developing in the 19th and early 20th centuries, strikers displayed a degree of heroism because they knew well that they and their families would suffer terrible hardships.[1] In more modern times strikers and their families could expect to have from various sources (including expenditure of some of their own savings) as much as 75 per cent or 80 per cent of their normal incomes available to spend during a strike,[2] though this percentage might be lower today due to changes in legislation affecting strike pay, supplementary benefits and tax rebate claims (see pages 54 and 60 below). Single strikers, not being eligible for supplementary benefit, may suffer more than married men, but it is rare now for a striker or his family to go cold or hungry. Their problem is that they may have to pay heavily later, not only for having consumed their savings but also in less predictable ways such as short time or redundancy due to loss of markets for their products. Even if they do get higher wages and extra overtime to catch up on lost production it may take them years to make up what they lost by striking, if they ever do. They 'strike now and pay later'.

John Gennard has done some superb pioneering research into the cost of strikes,[3] some of which is examined in more detail later in this chapter, especially in the table on page 54. His work has the distinction

TABLE 7.1 *Sources of income to strike householders*

	POST OFFICE	CHRYSLER	DUNLOP	CHAMBERLAIN
Strikers	Postal workers	Electricians	Clerical	Skilled manual
Date	Jan.–Mar.71	Aug.–Nov.73	Apr.–May 75	Jan.–Feb.77
Duration	47 days	98 days	20 days	20 days
Income during strike (all sources) [a]	£161.16	£665.40	£167.34	£185.89
Breakdown of income received	%	%	%	%
From the State				
Supplementary benefit[b]	14.5	1.4	–	2.6
Income tax rebate [c]	–	13.7	–	14.0
From the union				
Strikepay	–	21.9	–	–
Hardship pay [d]	1.6	–	0.2	–
From others				
Gifts, free housing etc.	2.4	0.9	1.2	0.9
Othersources	3.0	1.8	–	–
Earnings during strike				
Part-timejobs	3.0	1.9	1.7	3.4
Spouses earnings	15.6	11.8	54.5	11.0
At expense of striker				
Pre-strike pay in hand[e]	18.1	9.8	–	25.3
Savings spend	32.3	28.6	35.9	25.5
Borrowing	3.6	4.1	3.7	13.2
Deferred rent or HP	6.5	4.1	2.7	4.0
TOTAL (round off)	100	100	100	100

NOTES

[a]This total coming in during the strike or made available from savings etc. amounted to between 75% (Postal Workers) and 80% (Chamberlain) of the pre-strike family income though some individual groups had less than this (e.g. the female postal workers lived on only 62% of normal income).

[b]After the 1980 Social Security (No. 2) Act, supplementary benefit would have been reduced by a minimum of £12 per week for strike pay 'deemed' to be paid, whether paid or not. (raised to £13 in 1981)

[c]From July 1982, strikers were not eligible for their tax refund until after the end of the strike. If these four strikes had occurred today, therefore, the percentages of normal income during a strike would be considerably lower than quoted in (a).

[d]Hardship pay was from a union fund, to strikers in need who were not entitled to supplementary benefit (i.e. no family commitments).

[e]See page 58.

of being used and commended by both Management and Trade Unions as their main source of data in this very complex subject.[4]

The cost to companies is even harder to assess and the CBI[5] have examined various means of measurement but concluded that none were satisfactory. This is further discussed later in the chapter. Even harder to document are the longer term costs to the community and to the nation's economy, the lay-offs, the knock-on effects on other firms and on future confidence of British reliability in overseas markets. Nor is it any easier to assess the cost of public service strikes since these not only hamper the production of industry but also affect innumerable individuals who are thereby prevented from doing their best work in whatever is their job.

Nevertheless, even without precise figures, it is possible to assess the ways in which industrial disputes do cause short term and long term suffering to strikers and their families, to other workers and to the community at large.

HOW STRIKERS' FAMILIES LIVE DURING A STRIKE

One of John Gennard's most valuable studies was of four major strikes, between 1971 and 1977[6] in which he exploded the myth that strikers live primarily on social security and strike pay. While no two strikers are alike, his four examples were of sufficient variety (encompassing private industry and public service, skilled and unskilled manual workers, clerical workers, and high and low paid) to give the best detailed guide available as to how strikers' families live during a strike. He interviewed 8000 strikers and supplemented what they told him with data obtained from the Department of Employment (D of E) and the Department of Health and Social Security (DHSS). He found that between 42 per cent and 60 per cent of what they and their families lived on was at their own irrecoverable expense (e.g. savings and pay in hand) and that, if their own casual work and their spouses' earnings were included, they provided, in the lowest case 60 per cent and in the highest case an astounding 98.5 per cent from their own pockets, most of which they had to pay for by foregoing things they could have expected to enjoy in the months or years to come. By these means they cushioned themselves from hardship during the strike. Because people in industrial areas can now see for themselves that strikers and their families no longer suffer serious hardship *during* a strike, they are more ready to respond to strike calls from their own union officials or shop stewards without realizing how heavily they will pay later.

Table 7.1 summarizes some of the figures from Gennard's study of these four strikes,[7] all of which lasted long enough for strikers with families to become eligible for supplementary benefit. In the case of the Dunlop clerical workers strike[8] the amount claimed was a negligible percentage as most of the strikers, both male and female, had spouses who continued to earn during the strike. In two other cases the percentages were only 1.4 per cent and 2.6 per cent. Only in one case, in the prolonged Post Office strike by generally low paid workers, was the figure a substantial one (14.5 per cent). In the same study, however, Gennard quotes DofE and DHSS statistics which show that, between 1970 and 1977 only 22.8 per cent of strikers eligible for supplementary benefit actually claimed it. The highest percentage was 36.8 per cent in 1971 due largely to the number of postal workers who claimed it in that year. The lowest percentage was in 1977 (8.9 per cent) when only 10 000 out of 113 000 eligible strikers claimed their due. One reason for this is, of course, lack of awareness of their eligibility, but it is also yet another indication that, *during the strike itself*, strikers and their families were not as hard pressed financially as was generally believed, or more of them would have been driven to claim the benefits to which they were entitled. Another indication of this is that unions often withheld strike pay (by common consent) until after the strike was over so that the striker would not have his entitlement to supplementary benefit reduced. This, however, is unlikely to occur in the future because the supplementary benefit is now subject to deduction of the 'deemed' payment of £13 strike pay whether it is in fact paid or not.

Income tax rebate provided a substantial percentage of strike income in two of Gennard's cases (13.7 per cent and 14.0 per cent). This accrued by virtue of the fact that weekly PAYE deductions for the weeks prior to the strike had been based on the assumption that the taxpayer would receive full pay throughout the year so the employer had been required to deduct one fifty-second of this total tax each week. When wages ceased, the wage-earner therefore became eligible for a refund. There is a myth that the rebate is substantially greater if the strike occurs at the end of the tax year (i.e. in February or March). This is not generally so because the Inland Revenue calculates on the basis that wages will continue unchanged after the strike, so the rebate is very much the same whether the strike is early or late in the year.[9]

The payment of rebate did not begin, however, until the wages had actually ceased. Due to the 'pay in hand' system (see below) this was not usually until the strike had been going for two weeks. When the clerical workers themselves were on strike (as in the case of the Dunlop and postal workers' strikes) there were no wages clerks to pay out the tax rebate.

In July 1982, new regulations were introduced which froze *all* income tax rebate until after a strike was over. This meant a further drain on strikers savings, borrowing and in-strike earnings against the expectation of tax rebate on the return to work. This tax rebate itself has very substantially increased as a result of inflation and of the increasing number of manual workers whose earnings now enter the tax bracket. Thus, when the rebate freeze was introduced in July 1982 the average manual worker's earnings would have entitled him to about £30 a week rebate compared with £11–12 in 1978. The tax rebate now probably constitutes a larger percentage of total strike income, albeit not paid until after the strike is over.

Tax rebate and supplementary benefit were, and still are, the only sources of strikers in-strike or post-strike income paid by state. In the Dunlop strike, for the reasons given above, they amounted to nil. In the other three they totalled 14.5 per cent (Post Office), 15.1 per cent (Chrysler) and 16.6 per cent (Chamberlain) of the strikers' income. These might today amount to much the same percentages overall, since the reduction (by 'deemed' strike pay) of the supplementary benefit would be offset by the higher tax rebate, though more of this percentage would be post-strike rather than in-strike. Both before and since these two changes were introduced, however, the 'state subsidy' of strikers' income was and remains a relatively minor factor.

Turning to the contribution from the union, this too was generally small in the four strikes analysed by Gennard. Only in one case (Chrysler) was there any strike pay at all, though in two of the cases some strikers, not eligible for supplementary benefit, received small amounts of hardship pay. The Chamberlain strikers received nothing.

Gifts from friends and relations, mainly in the form of free board and lodging or excusal from paying rent, made a substantial difference for some individuals though the overall percentage was small.

Spouses' earnings provided more than half the emoluments of the Dunlop clerical strikers and over 10 per cent in all the other cases. Nevertheless, as recorded earlier, most of the income fell into the 'irrecoverable' category. Of this category, savings were the largest item. Only a very small minority of these were saved for the specific purpose of financing the strike (about one-fifth of the Chrysler electricians total savings and a negligible proportion in the other three cases). The bulk of the savings expended were those earmarked for such things as holidays, or purchase of cars or washing machines, which had therefore to be foregone or postponed. A small amount was made up of borrowing (e.g. being given credit by local shops) and deferment of rent and hire purchase payments, landlords and HP firms generally being reasonably accommodating about this.

The other major source was 'pay in hand'. Most firms pay wages a week in arrears. This can best be illustrated by a hypothetical example. Suppose a strike begins on Friday, 5 March. The hours of work, bonuses, PAYE, etc. for the last week of work (1–5 March) are computed by the wages clerks during the following week and paid out on Friday, 12 March. Thus a worker who goes on strike from midnight on Friday, 5 March will have one week's pay in his pocket received that afternoon and will draw a further week's pay on Friday, 12 March.[10] The DHSS and others assume that the pay packet received on a Friday is for livelihood until the next one is due the following Friday, so the striker's income will not in fact be directly affected until Friday, 19 March – that is, two weeks after the strike begins.

It is, however, misleading to regard this as 'financing of the strikers by their employers'. The pay for these two weeks is in fact lost and will be reflected, in normal circumstances, by a delay of two weeks before the first pay packet is received after resumption of work. Employers do, however, sometimes make an immediate advance of pay, as a gesture of conciliation, when strikers resume work, though this normally remains a debt to be deducted from future pay packets. If, of course, the employer, as part of the settlement, were to waive this debt the advance would become a bonus. In this case the employer would have helped to finance the strike but this is relatively rare and the striker would certainly not have been able to rely on such a bonus when he began the strike.

The striker, however, may well feel able to rely on the prospect of the company calling on him for overtime work after the strike to make up for lost production and fulfil market commitments. Since this overtime will be paid at premium rates, the company can in this respect be regarded as helping to finance the strike in retrospect, albeit unwillingly.

It has been observed that, in times of recession, strikes over minor issues are fewer than normal, presumably because the strikers fear that those involved might be subject to management reprisals, but that workers may be more willing to join major stoppages; also that these major strikes tend to last longer. One explanation of this is that the workers may feel that full production would in any case lead to layoffs and short time working, and that the management may be happy enough not to pay wages for a period knowing that they can make up the lost production later. In other words, for both workers and management such strikes are a form of work-spreading.[11]

Against this, the experience of 1981–82 suggests that, in a really deep recession and in face of high unemployment, workers may fear large scale redundancies, cutbacks or even the risk of driving their employer into bankruptcy. There are also the influences arising from the Employment Act 1980, including the 'deeming' of strike pay and the employer's power of selective dismissal of strikers who do not return to work by a stated date. Both of these pressures applied to the train drivers in the ASLEF strike in July 1982, because their union gave no strike pay and British Rail did name such a date. It was probably the realization that the strike would collapse before that date which led the TUC to urge ASLEF to settle on British Rail's terms.

Another source of cushioning, which did not apply in any of the four strikes in John Gennard's study, is holiday pay. If the strikers start their strike at the moment when they are due for the advance of pay normally given when a paid holiday is due, they can receive this advance on the same Friday as their last pay packet. In two of the strikes we ourselves studied (not John Gennard's) this was a factor. In one case (see Chapter 9) the strike began as the firm closed for the short Christmas holiday so the last pay packet included an extra £40. The firm then deducted this from the weeks arrears of 'pay in hand'. Though it was, of course, included in the first pay packet after the strike, the deduction greatly embittered the strikers. In the other case (in the motor industry) the strike began at the start of the three week summer holiday in July/August, so the strikers started the strike with four weeks' pay in their pockets. This gave them plenty of time to plan how to sustain themselves before they became short of cash. They expected that, realizing this, the company would make an offer to settle the dispute before the holiday. They did not do this, but the advantage of this cushion enabled the strikers to hold out for a 13 week strike and the company resolved to time future negotiations so that they would not reach a head as a holiday period approached.

'STRIKE NOW – PAY LATER'

We have already seen the way in which, although a striker and his family may have 75–80 per cent of their normal income during a strike, over half of this will eventually have to be paid for later.

In some cases this may well seem worthwhile. In the 1977 firemen's strike, for example, the average striker sacrificed £500[12] but under the settlement of the strike the firemen received a guarantee that their wage would in future be maintained by a review body with the top 25 per cent of the average manual wage.

If, however, industrial disruption is continuous, as it was in the motor industry in the 1970s, these losses may be a running sore. A car worker at BL's Cowley plant in April 1974 reported that he had received only 22 full pay packets out of a possible 53 in the past financial year, that frequent lay-offs had reduced his average working week to 33 hours and that he had lost on average £10 each week (this at 1982 prices would be at least £20 a week or £1000 per year). This scourge of strikes in the motor industry and its partial alleviation in the 1980s is discussed in Chapter 10.

In the 9-week stoppage at Ford in autumn 1978, one shop steward reported losing £55 per week or £500 in all[13] and the average was £300, which it was estimated would take six months to make up from the wage rise achieved including the probable estimate of extra overtime pay.[14]

The trade unions lost even more heavily and this, indirectly, was a loss to their members. In the 1978 Ford strike the trade unions paid out £4 million in strike pay, of which £1m was paid by the AUEW, one of whose senior officials estimated that it would take the 12 000 members involved in the strike seven years to replace. In practice, union members pay for this either by receiving a reduced service or by increased contributions later.

In the 1979 national engineering workers strike the AUEW paid out a further £2 million; and in the 1980 steel strike they paid out another £1.3m in strike pay at £9 per week.[15] The main union involved in the steel strike, the ISTC, spent £2 million in publicity and picketing but provided no strike pay, so the individual steel workers lost an average of £1000 each which, on the basis of the settlement, would take them two years to recover.[16] Their greater loss, however, was in terms of loss of markets and redundancies. Between September 1979 (just before the strike) and 1981, 74 600 out of a workforce of 184 600 were made redundant and the BSC target was reduced to 90 000 – a cut of more than half.[17]

In one dispute an attempt was made to calculate the effect on both sides – a stoppage in the printing industry in 1979. It was calculated that it would take five years for the wage increase conceded to make up the wages lost in the strike; but that the cost of that concession, if not passed on, would reduce the employers' profit margin to 1 ¼ per cent of turnover.[18] In the event they did have to pass on the costs with the result that the British printing industry has become uncompetitive, most printing is now done overseas, and the printing workers have lost proportionately more jobs than almost any other. In an excellent analysis written at the time, Cyriax and Oakeshott prophetically likened this result to a long and punishing game in which each side scored equally 'but each side scored through its own goal'.[19]

Less tangible is the cost *to workers* of the deferment of modernization of their plant due to the loss of anticipated company profits and – even more permanently – to the diversion by multinational companies of production to other countries less disrupted by labour disputes. After the 1978 Ford strike, new machine tool projects worth £400 million were postponed resulting, not only in declining prospects for British Ford workers, but also loss of orders for the British machine tool manufacturers. Another example is given in Chapter 9, where a dispute involving 100 workers resulted in the diversion of their particular production to the French subsidiary of the same corporation, and the loss of one-third of their share of the European market for their product, contributing thereby to 2500 redundancies.

THE PRICE PAID BY OTHER WORKERS

This is one example of the most tragic knock-on effect of strikes – the effect on other workers in the same or associated industries. The four week strike by TGWU lorry drivers in 1979 caused 235 000 industrial workers to be laid off. The national AUEW strike later that year caused 60 000 lay offs. The Dunlop clerical workers strike (see above) is also examined further in Chapter 8, where it will be seen that a strike by

660 clerical workers led to the laying off of over 2100 Dunlop manual workers (not in dispute) and of 6000 BL workers (not in dispute) and a huge loss of funds for investment and modernization of these two British industries.

The coal strike of 1974 reduced the whole of British industry to a three-day week. Though surprisingly little production was lost because shift and piece workers were anxious to earn as much as they could in the three days available for them,[20] the price paid in lost earnings by the nation as a whole was horrific.

The long term cost of strikes to the strikers themselves, and to their fellow-workers in other industries dependent upon them, is inextricably bound up with the effect of the disputes, direct and indirect, on the performance of British industry, on its modernization and on its markets, on the British balance of payments and in a myriad of ways on the community as a whole.

THE COST TO FIRMS HIT BY A STRIKE

Sir Alex Jarratt, formerly a senior civil servant and one time secretary to the National Board for Prices and Incomes, now Chairman of Reed International, has estimated that strikes cost employers about 100 times more than they cost trade unions.[21] These losses are a major cause of the British disease. It was recorded in Chapter 1, for example, that the capital per worker in German industry is double that in British industry.[22] It should therefore be no surprise that the German worker produces so much more per man-hour than the British worker, as a man with a plough tills more ground than a man with a fork. Some of the responsibility for this falls on successive governments (for failing to give sufficient tax-incentives for investment), on British financial institutions (for investing abroad rather than in UK), on management (for their reluctance to invest risk-capital) and, of course, on the multinational corporations which provide such a large proportion of the capital for British industry. All except the first of these, however, can be ascribed at least in part to the poor record of return on capital investment in Britain which is to a large extent due to the high level of industrial disruption; also to the fact that every £100 million lost in profit for an industry is a £100 million less available for investment.

The most expensive strike in recent years was the national AUEW strike in 1979 which cost British industry £200 million in lost sales.[23]

The 1980 Steel Strike cost the BSC £130 million in lost orders.[24] Assuming that the profit on these sales would have been about 10 per cent the loss of funds available for investment in British industry in the 10-week engineering strike would have been £20 million, and to BSC in the 13-week steel strike £13 million. Nothing can replace this loss, especially as a large percentage of the orders were for export, and many of the remainder will have been replaced by imports. The loss will mainly be paid for in loss of investment, thence lack of modernization, loss of competitiveness and therefore loss of earnings and jobs.

A simple calculation of loss of production can, however, give a false measure – too large or too small – of the cost of an industrial dispute. A motor manufacturer, for example, may have a stock of 5 000 cars in his parking lot, each one priced at £5000. If his production capacity is 1000 cars per week, a five week total strike, far from costing him £5 million a week (£25 million) in lost sales, may actually save him money as he will be relieved of the bill for wages and lay-off pay while the stock is cleared. Trade unions have frequently accused managements of precipitating strikes, or deliberately prolonging them, for this purpose. Another similar accusation is that, to avoid unprofitable under-utilization of plant, managements build up surplus stocks in order to strengthen their hands for a later confrontation with the trade unions over wages, knowing that they run no financial risk if it leads to a strike, whereas the unions and workers may lose heavily.

The CBI made an attempt[25] in 1981 to assess the reliability of various ways of evaluating the cost of strikes to the employer. The simplest and most commonly used method was to measure the lost production by multiplying the average output per day by the number of working days lost. They pointed out several shortcomings, however: some of the lost production may later be made up (though partly at the cost of extra overtime pay); some of the intended output might not have been sold; and many of the normal producton costs – wages, materials, components, fuel and power – may be saved during the strike.

The CBI next considered the 'value-added' approach. This is based on the wages and salaries foregone by the strikers, with additions for the non-labour element in value-added per worker, plus, if appropriate, the interest cost of the delayed inflow of cash arising from delay in fulfilling orders. It is concluded that this method is best and quickest for costing smaller scale industrial action such as overtime bans or token strikes or strikes by a proportion of workers which do not bring production to a halt. On the other hand, its shortcoming is that it assumes that value-added once lost cannot be recovered and that the cost of losses is directly proportional to the number of worker-hours lost, neither of which is generally true.

The CBI next examined the measurement of the cost of a strike by the gross value of sales lost. This method does overcome the misleading factor caused by stockpiles (as in the 5000-car example above), but it too can give misleading results. The manufacturer's mark-up values will be based on an anticipated level of sales and, if this target cannot be reached because of lost production, he cannot recover the fixed costs which would have been borne from the missing sales. Unlike the striker who has over-paid PAYE tax, the employer has no *deus et machina* to give him a rebate after the strike is over.

Another method is to estimate loss of profit – i.e. the difference between the actual profit achieved over the period and the forecast profit if there had been no stoppage. This, however, depends on the accuracy of the forecast profit, which may be vulnerable to many unpredictable factors other than strikes.

INDIRECT AND LONG-TERM EFFECTS ON BRITISH INDUSTRY

Over half the major stoppages in strike prone industries result in layoffs of workers in other connected industries which supply them with the material and components or which rely on their products for their own production, as described above. This also has a damaging effect on efficient management. A component firm deprived of outlets may be faced with heavy storage costs for its unwanted product or with cut-backs or cessation of production. A manufacturer starved of components will at the best be deprived of flexibility as his stockpiles run down, and may often have to cease production. Examples of all of these phenomena are examined in the case studies in subsequent chapters.

Direct losses, however large, are likely in the end to be exceeded by the indirect and longer term effects, not only of the stoppages themselves, but of the many industrial conflicts resolved without strikes, and of the economic damage done by firms giving way to avoid the dispute ever coming to a head. One manager of a British subsidiary of a large multinational told us that as soon as there was any threat of a dispute the Corporate Headquarters in USA immediately said 'pay them whatever you have to but keep production flowing' – and then cut back on its activity in Britain. It is no surprise that the Corporation has now ceased all activity in Britain and had serious financial difficulties in USA.

A number of writers have produced similar evidence that strikes themselves have little direct effect on wage inflation but that the constant *threat* of industrial action is the prime mover in cost-push inflation.[26] Because wages in strike-prone industries (mainly mining, transport and motor manufacture), are driven up by this threat, wages in other industries in which the trade unions do not have the power to exert a convincing threat can get their wages raised by appealing for equity (the leapfrog process). High wages and low productivity produce lack of competitiveness and encourage rival firms to increase their market share (for an example see Chapter 9). Investors, particularly the big financial institutions and merchant banks, lose confidence and divert to more reliable investments. Efficient and innovative managers get tired of the hassle and accept the lures of headhunters – the best seldom have trouble in getting a more agreeable or rewarding job. For the same reason, the firm's image drives away the best graduates. The same applies to many of the best workers who get sick of continuous interruption of their earnings (as in the example of the Cowley workers on page 60 above).

The psychological effect on British managers was well put in the Donovan Report:

> Some managements lack confidence that the plans they make and the decisions they reach can be implemented rapidly or efficiently or, in extreme cases, at all. . . . The country can ill afford the crippling effect which such management attitudes are liable to have on the pace of innovation and technological advance in industry.[27]

The Personnel Director of one of the large motor corporations told us that this constant conflict breeds fire-fighting managers who do no more than react to problems rather than attempt progressive change, and who become attuned to being inefficient due to the strike propensity of their industry. Many senior managers told us that the most serious problem of all is the diversion of management time and effort to the avoidance and resolution of industrial conflict. In the middle 1970s managers in the British motor industry estimated that about 50 per cent of their time was spent in this way compared with 5–10 per cent amongst their opposite numbers in Belguim and West Germany[28] (see Chapter 10).

Closely following from this syndrome is the excessively cautious management which results from the constant worry of industrial disputes, including those which are resolved after days and nights of exhausting negotiation. To avoid this continuing agony, managers may be slow to react to change in demand or introduction of new technology, knowing that this will lead to a prolonged series of fudges while their industrial engineers hammer out norms for the different new processes with shop stewards determined to establish a slow rate with a view to acquiring large bonuses by 'beating' that norm as soon as the rate for the job has been accepted or piecework introduced on that basis. An example of such a process, and of its devastating consequence of loss of markets and therefore loss of jobs, is described in the case study in Chapter 9. Faced with the prospect of such a struggle the management may settle for the easy life, quietly balancing their books with safe established products – until these are overtaken by the new products developed by rivals in Germany and Japan. This is a disastrous recipe for meeting the challenge of the era of change that lies ahead.

For the same reasons, managements may be reluctant to bid for attractive orders if they fear that restrictive practices and blackmail to reach hard bargains for expanding output will make them unprofitable. They may reject potentially profitable investment, knowing that this means innovation or expanded production or both, which will surely launch yet another exhausting industrial dispute.[29] The tragedy is that, while this kind of conflict gives ulcers to the managers, it also in the end produces unemployment on the shop floor.

An even more damaging but less tangible effect of industrial disputes is loss of markets, especially to foreign competitors. Delays in delivery cause disruption and expense to customers (see the example in Chapter 9) and perhaps layoffs in their own factories. They will be tempted to divert future orders elsewhere, or at least a part of them so that they become less vulnerable to the unreliability of a single supplier. A reputation for strike-proneness is likely to make it harder to win new markets. Moreover, every delay in delivery which is blamed on industrial disputes (sometimes unscrupulously by exporters trying to excuse late deliveries resulting from other causes) reflects on British industry as a whole, thereby turning foreign buyers away from efficient firms which are free of strikes and have a record of prompt delivery.[30]

THE EFFECT ON THE COMMUNITY

All of these things add up to a lower income per head for the British people, less than 60 per cent of that of their German counterparts, and an even gloomier prospect for the future as our competitors' modernization leaves our own industries further behind. Parallel to this, the adverse balance of payments arising from stultified exports and from imports substituting for our own shortfall in production can only be met by squandering our one major natural resource – oil – and when that runs out, the nation we bequeath to our children will be faced with bankruptcy, like the profligate military dictatorships of South America and the ailing Communist states of East Europe.

The low rate of production of real wealth per head in Britain is also reflected in the availability of funds for the exchequer from direct and indirect taxation, corporation tax and national insurance. This can only mean poorer public services. Though the British people are amongst the most highly taxed in the world, they have some of the most impoverished public services, because even high taxes on low earnings may give the government less to spend; put another way, 30 per cent tax on £600 is less than 20 per cent tax on £1000.

The inconvenience and hardship inflicted on the community at large by strikes is even less susceptible to accurate measurement than the other ingredients of the price we pay. Strikes in private industry (or in publicly owned industries like BL cars) result mainly in people having to accept delays in replacement or repair of their consumer durables, apart from the wider economic costs already described. It is a disgrace, but a well-deserved one, that in an industrial country urgently needing work for its factories, over half the cars and a huge majority of the washing machines, dishwashers, television sets and motor cycles have to be imported because of consumer exasperation with the value and reliability of our own products.

One of the most pernicious effects of the workings of our industrial relations system is that the redundancies resulting from the sapping of industry by industrial disputes are themselves the cause of further industrial disputes which are usually resolved, after prolonged bargaining and perhaps more strikes with the magic formula of 'natural wastage'. This means that ageing workers and volunteers for redundancy retire without replacement. The inevitable result of this is youth unemployment, amounting in some cities to 50 per cent, and often higher for young black school leavers. It would be hard to find a surer recipe for producing a demoralized, frustrated and embittered generation for the future and this was a major cause of the explosion of riots in British cities in July 1981.

The greatest personal inconvenience and hardship for members of the public, however, arises not from strikes in industry but from strikes in public services like transport, energy and hospital services. These hardships are almost impossible to measure. How can we cost the suffering of a patient who cannot get hospital treatment, of a dying mother whose son cannot reach her, of a father prevented from earning a living for his family or of a young man who is desperate to pursue his love for a girl in another city and cannot get there?

As was pointed out in Chapters 4 and 5, public service strikes are severely restricted by law or banned altogether in the USA and in most European countries. There is little doubt that the great majority of the British public – including the majority of public service workers themselves – would vote for a similar policy in Britain, provided that it included safeguards (e.g. a review body) to guarantee such public service workers a fair wage not inferior to that of equivalent workers in other industries where pressure can be applied by strikes (see Chapters 11 and 18).

The means of doing this, along with the means of attacking the self defeating diseases in British industry, is discussed in the final chapters of this book. Excessive or repressive legislation could so poison industrial relations that the losses from non-cooperation and apathy could lead to even greater evils. The answer must lie in greater participation in management by directly elected workers representatives, by better communication of what is really at stake in low or disrupted production, and thereby the mobilization of a work force which knows that prosperity depends on high producton at a competitive cost for the home and world markets.

8 Dunlop: the Knock-On Effects of a Strike

A VULNERABLE INDUSTRY

The engineering industry is notoriously vulnerable to industrial disputes and none more so than the motor industry with its interdependence of the big motor manufacturers and the numerous specialist component manufacturers. A major stoppage at Ford or BL can have a disastrous effect on the cash flow of a firm making, say, carburettors, speedometers or tyres. Similarly, a stoppage in one of the component firms may not only halt production in the giant assembly plants, but thereby also halt the demand for the products of all the other component firms as well. Coming down the scale, a strike of a small key group within a factory – e.g. toolmakers, electricians or clerical workers – can cause massive layoffs in their own factory and thence in all the others, big and small.

As in every other dispute worth the name, both sides usually have a case and feel, or may be led to feel, that a vital principle is at stake. The particular group of workers involved may believe that they are unjustly deprived of their fair reward in relation to others and that unless they stand on principle they will be left permanently behind and that their families will suffer. On the other hand the management of a small firm, with their account books staring them in the face, know that if they unleash a wage hike they will be faced either with raising their prices and losing their market to their competitors or selling at a loss. In either case the result would be bankruptcy and then the whole work force would be on the streets.

When a large component firm is involved the damage can be catastrophic because it is likely to affect several if not all the major motor manufacturers and this in turn hits the whole industry. The example chosen for this case study is the Dunlop clerical workers strike in the spring of 1975. The figures for strikers incomes were given, in comparison with those in other strikes, in the Chapter 7. This chapter examines the strike itself and its effects on all concerned.

THE STRUCTURE OF THE FIRM

The Dunlop Engineering Group in the West Midlands in 1975 employed a total work force of 5000 of whom 660 were clerical workers. The Group was divided into four divisions; Aviation Division, Wheel Division, Suspension Division and Engineering Industrial Products Division. The Aviation Division supplied the whole of the British aircraft industry. The Wheel Division supplied 80 per cent of the British car industry (including BL, Talbot and Vauxhall) and turned out from 150 000 to 200 000 wheels per week. Wheels are bulky so the motor manufacturers can only keep small stocks. A stoppage at Dunlop would stop BL car production within a week. There was no alternative supplier in Britain who could provide the quantity at short notice so if motor manufacturers were frequently starved of wheels they would turn to Europe. Once lost, markets would be hard to recapture. Since motor manufacturers insist on long term contracts and prices are highly competitive within the EEC, Dunlop would pay heavily for either serious stoppages or price rises and loss of contracts would mean loss of jobs.

Management were well aware of this. So, too, were most of the trade union officials and shop stewards. It may well be, however, that the leaders of the small number of clerical workers in the firm believed, and persuaded their members, that this actually strengthened their hand, and that the firm would meet their demands rather than risk serious disruption; in other words, that they had powerful leverage.

THE STRIKE

The Clerical Workers dispute came to a head in April 1975. There were two unions involved – APEX and ACTSS (the white collar section of the TGWU). There were only 660 clerical workers involved (500 of them female) and they had, ironically, the reputation of being the least militant workers in the motor industry. They did, however, wield considerable power since they handled the essential bills of lading so they could stop all movement of goods into and out of their factories.

Their claim was initially for a rise of £15 per week which was lowered to £10 ten days after the strike began. Their aim was to close the gap to match the £40 a week minimum clerical rate in other Coventry engineering plants such as BL and Massey Ferguson. To put this claim in perspective the wage of the lowest paid workers in Britain in mid 1975 was about £25 per week, the average manual workers wage about £50. The rate of inflation was 25 per cent and – to everyone's alarm – on a very fast rising curve (20 per cent in February rising to 27 per cent in August).[1] Dunlop feared that a sudden rise of £10 a week (33 per cent for those on £30) would lead to similar claims throughout the firm so they restricted their offer to a range of £7 to £8 for various grades of clerical worker. The clerical workers refused this and went on strike on 18 April 1975. The dispute was eventually settled with increases of £8.51 for males and £7.35 for females – which was very much the same as the management offer of £7.28 on 21 May.

Though the strike lasted less than 4 weeks, production was brought to a complete halt within a week. Some 2100 Dunlop manual workers were laid off – over half of them in the Aviation Division. This was three times as many as the clerical workers themselves and they all lost three weeks or in some cases four weeks' work.

TABLE 8.1 *Knock-on effect of clerical workers' strike*

Dates	*Clerical workers*		*Layoffs (manual)*		*Short time (manual)*	
	Number on strike	*Hours lost*	*Number laid off*	*Hours lost*	*Number on short time*	*Hours lost*
18–25 April	660	28,000	194	5,048	525	14,810
26 April—23 May	660	77,700	2,132	251,806	55	2,200
Total		105,700		256,854		17,010

The effect on the rest of the motor industry was devastating, because motor manufacturers had cut back their stocks of wheels to improve cash flow. BL was particularly hard hit. Mini and Allegro production was halted, and output of the Marina, Triumph and other models was cut back. Altogether in Coventry, Cowley and Birmingham (Longbridge and Castle Bromwich) over 6000 men were laid off (three times as many as at Dunlop itself), 1.4 million man hours lost (five times as many), and the equivalent of production of 23 000 vehicles lost. The daily loss of production at BL was 1400 cars valued at £1½ million per day.

These figures, in fact, understate the case because about half the man-hours lost in Dunlop were in the Aviation Division and, since aircraft completion schedules are much longer than for the cars and buffer stocks were adequate, the strike had little effect on the aircraft industry. Thus the losses of man hours by layoffs in BL were 10 times those in the Wheel and Suspension Divisions at Dunlop and about 25 times those of the striking clerical workers in those divisions. The other manufacturers dependent on Dunlop (Chrysler and Vauxhall) were less hard hit because they were able to get some wheels from GKN and Rubery Owen, while Ford were unaffected because they manufactured their own wheels, but the effect on BL was catastrophic and will have contributed not only to their continuous and heavy losses (which had to be picked up by the British taxpayer); also to the decline in BL's share of the home car market from 32.7 per cent[2] in 1974 to 17.1 per cent[3] in 1980 and to the consequent decline in BL's workforce from 197 000 in 1978 to 129 000 in 1980[4] and to 87 000 in 1982.[5]

The loss to individual manual workers in Dunlop was very heavy since national agreements on layoff pay exclude layoffs due to a dispute of other workers in the same company. Thus, while BL workers will have received layoff pay for some days, Dunlop manual workers received none. It is probable that few, if any, had much sympathy with the clerical workers. Most of them resented being deprived of their livelihood, and were aware of the likely long term effects on their jobs but were powerless to do anything about it.

THE AFTERMATH

The clerical workers themselves gained little from the strike in the short term and lost heavily in the long term. The 660 clerical workers in Dunlop had fallen to 448 by 1981 and their overall average earnings in real terms fell by 25 per cent.[6]

The only redeeming feature from the strike was that it jerked Dunlop management and trades union officials into a major review of their industrial relations procedures. By joint wage structure agreements signed in 1976 the number of job rates was reduced from 493 (in 1971) to 7, by grouping large numbers of trades into 7 grades, making negotiation much easier. Job evaluation is now done by a team of two managers and two convenors, who investigate grievances and handle appeals. In practice they almost invariably arrive at a unanimous decision.

At the same time, the communications system between management and labour has been vastly improved. Each Division of the Dunlop

Engineering Group has a Business Review Meeting every two months on average and sometimes monthly. These are chaired by the General Manager and attended by board members and representatives of all grades in the Division. These meetings concentrate solely on business performance and, in order to keep to this, separate meetings are held with Personnel Managers to deal with problems like catering and working conditions. The Divisional Business Planning Meetings are joined at least every three months by Group directors who brief the representatives on the Group situation as a whole in terms of marketing performance and trends. Discussion is frank and quickly develops a constructive attitude; how best can *we* keep our costs down in order to seize this market opportunity to keep earnings high and avoid redundancies. Provided that the representatives retain the confidence of their members on the shop floor and can communicate these facts of life to them, the loss of profits and earnings from industrial disputes should be kept to a minimum.

In practice this has worked. Dunlop have had no major disputes since 1976. That, however, cannot justify the 1975 strike any more than a subsequent improvement in fire precautions could justify arson. The damaging disputes from which no one gains and everyone loses can only be averted by better communication which is the subject of Part III of this book. This dispute was a striking example of an 'own goal' kicked on principle with eyes wide open.

9[1] Industrial Conflict in a Multinational Subsidiary

PURPOSE OF THIS CASE STUDY

A damaging strike in a tractor factory in the West Midlands has been selected for this case study because it shows how a minor dispute involving 100 men had a devastating and lasting effect on the firm and especially in subsequent layoffs and redundancies of the labour force. Both management and unions made mistakes and, as a minor consolation, both learned from these mistakes and enjoyed industrial peace for the subsequent 4½ years. This peace was then broken by another strike over redundancies which must at least partly be ascribed to the first strike.

The management and union representatives both gave the fullest and frankest cooperation in this case study, sometimes together and sometimes separately. Despite strong differences of opinion, their personal relations at working level remained excellent. The interpretation of the author and his research assistant of these conflicting views cannot reconcile them though it aims to reflect both as fairly as possible. Inevitably, one side or the other will disagree with different parts of this interpretation. For this reason, and out of respect for both sides frankness, the corporation concerned will be referred to under a pseudonym, the 'Atlantic Plant Corporation' (APC). This was not done at the request of the company but because the author preferred to do it this way.

This case study particularly underlines the problems of subsidiaries of foreign-based multinational corporations, which can freely switch production to other countries (in this case France) where better industrial relations may give prospects of more reliable and profitable production; or they can reduce investment or, in extreme cases, cut it off altogether. APC(UK) had this problem in common with all the major motor manufacturers except BL. Even BL, though underwritten almost entirely by the British taxpayer, has to contract for much of its manufacture in Europe and Japan in order to survive. All share the kind of problems covered by this case study (productivity, work values, piecework rates and rivalry between different assembly line gangs each wanting the same job and bonuses arising from it). Since APC(UK) operates on a smaller scale than the volume car industry it was easier to isolate this dispute from extraneous factors and therefore to draw valid lessons from it.

It should be placed on record that this is not the first time that APC(UK) have cooperated in this kind of research. Both management and union representatives have been a model in this respect, cooperating in several other industrial research projects so that the rest of the industry is in their debt for enabling their experience to be shared.

Writing the story of this dispute has been like writing a dramatic tragedy. Both sides, acting with sincerity and best intentions, felt that they must stand on principles which were vital to them, knowing that they (and in the case of the shop floor workers – their families) would pay a heavy price for doing so. At least one shop steward admitted that he had to avoid letting his members know what this would cost them, both in earnings and in future redundancies, because had they reflected on this he could never have carried them with him in fighting the principle. Both the firm and the unions agree that they lost heavily from fighting the dispute, but both felt that it would have been worse to give way. The real and lasting price for APC(UK) and its work force, however, is that their share of the UK market fell from 33 per cent to 20 per cent largely as a result of the dispute though there were other factors too. During the next four years 2500 manual workers lost their jobs, adding further to the already horrific unemployment rate of 15–16 per cent in the West Midlands in 1982, far above the national average and rising. Some of these jobs would have been lost in any case from the recession but nothing like so many, and that extra loss has been largely taken up by workers in other countries. Nothing could better underline the vital need for management and labour in Britain to join forces in achieving high productivity to produce a competitive product.

APC(UK) IN 1976–77

APC manufactures a wide range of industrial and agricultural machinery and their West Midlands factory is the largest tractor factory in Western Europe. Of APCs' world wide labour force of some 70 000, this factory in 1976 employed 7000 of whom 5000 were manual workers and 2000 staff. The factory had a capacity of 90 000 tractors or tractor sets (for assembly overseas) per year. In 1976 it produced 85 000 but by 1981 it was producing little over half its capacity (45 000) and still less by 1982, though since then they have begun to recover some of their market share.

APC(UK) operated a pre-entry closed shop which meant that, in effect the trade unions did the recruiting. Of the 5000 manual workers about 60 per cent belonged to the TGWU, 30 per cent to the AUEW, 8 per cent to the National Society of Metal Mechanics (NSMM) and the remaining 2 per cent to others such as the Electricians and Sheet Metal Workers' Unions.

The factory had a mature work force better paid than most, with a low turnover of less than 5 per cent per annum. Eighty per cent of the work force had been with the company for over 5 years, 25 per cent of them for 20 years or more. This was also reflected by union membership, 60 per cent of members being over 45 years old and only 10 per cent under 30 – a work force with heavy family commitments. Their natural reluctance to strike, however, was to some extent overridden by more than usually militant unions with a high strike record.

The number of shop stewards in the manual unions varied from 120 to 140, of whom 70–75 were TGWU and 50–55 AUEW with 4 or 5 NSMM and another 4 or 5 in the smaller unions. The average steward in the assembly shop (the scene of this strike) represented about 40 men, the numbers being lower amongst indirect workers and those in the craft areas and machine shops.

There were three convenors each elected annually, the AUEW convenor by the AUEW shop stewards and the TGWU and NSMM convenors by ballot of all their union members. The AUEW convenor was Chairman of the Joint Shop Stewards Committee (JSSC), and the TGWU convenor was its Secretary. The third convenor, from the NSMM, also represented the smaller unions. All three convenors were engaged full time on union duties but paid by the company at the rate equivalent to the top of the second quartile of the members they represented.

CAUSES OF THE 1977 DISPUTE

In the 1977 strike the crucial production units were the assembly gangs. Each operated an assembly track and was normally some 50 to 100 strong, led and organized by a shop steward or stewards whom they elected.

In the early 1970s the EEC had published mandatory Safety

Regulations which required all tractors sold in member countries to be fitted with integrated cabs by July 1976. Of APC(UK)'s production of 85 000 tractors and tractor sets in 1976, 90 per cent were exported, some to Europe and many more to The Third World and other countries unaffected by EEC Regulations.

The company therefore signed a three year contract for delivery of these cabs by GKN Sankey starting in 1976, and put in motion discussions with the shop stewards of the two new groups to design the necessary manufacturing processes and to arrive at work rates for the tasks involved. These gangs had established the principle of considerable autonomy, exercised by their shop stewards, who actually controlled the tasks and hours of work and therefore the earnings of individual members of the gangs, and also their shifts (where appropriate), holidays, layoffs and manning levels (jointly with management).

The plan was to operate the two assembly tracks (H and Y) already in operation to make tractors for export outside Europe (i.e. not requiring integrated cabs) and to reactivate a third track, previously moth-balled – Track G – to build the integrated cab tractor. To feed the cabs to G Track was a fourth track (Z) on which the cabs would be trimmed (i.e. fitted with seats, instrument panels etc.). (See Figure 9.1.)

For this purpose, two new gangs were to be formed, each with about 50 men:

Gang 18 would trim the cabs on Z track

Gang 17 would receive the cabs on G track, fit them onto the chassis as it moved down the track and build the complete cab tractors including wheels, batteries, radiators etc

Two problems at once aroused controversy between unions and management and also between rival gang shop stewards. They had found it hard to agree on the actual tasks of gang members, manning levels and rates-for-jobs. The assembly lines were not on piecework (NOP) at this stage but on static earning with no incentive bonuses. A problem also arose in that the two large gangs each 100 strong (gangs 5 and 6F) already operating the two non-cab tracks H and Y felt that they should operate G track, because their work was currently stagnant and had exhausted their potential for earning bonuses.

It is necessary at this stage to describe the procedure whereby new production processes were introduced and how these could be developed to a piecework system.

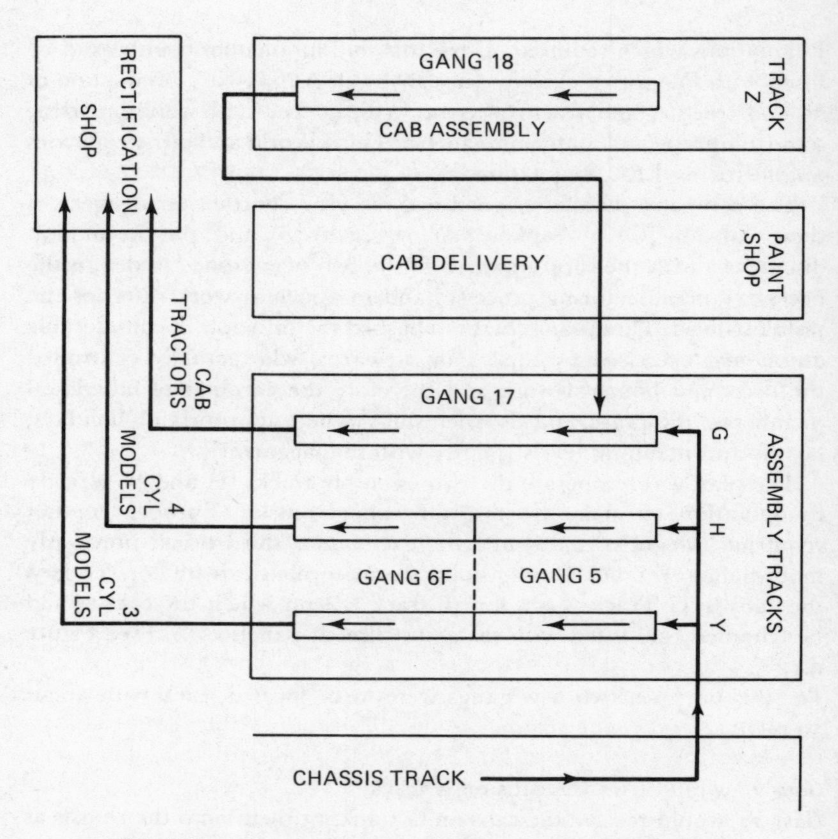

FIGURE 9.1 *Assembly of cab tractors, APC (UK) 1977*

Management would first outline the routeing of the material onto the track and set out the processes which had to be completed to build the tractor. The gang shop stewards would then look at this routeing, consider the layout of the track, the positioning of tools, congestion problems etc. and do their own stop-watch timing of each operation. From this they would formulate a proposal for the size of the gang required and an appropriate production norm (e.g. four tractors an hour). It was standard practice to accept that there would be a learning curve as the job progressed and that work values and rates for the job would be adjusted accordingly over a period. Thereafter, once the process had got into its stride, it would be possible to change to piecework which would, in the end, be beneficial both to the company and to the process workers.

Here, a conflict in tactics would arise. The shop stewards would try to build a case for as large a gang as possible, to create jobs and, later on, once norms had been established and piecework introduced, to shed jobs from the gang in order to claim increased productivity and therefore higher earnings for those remaining in the gang. On the other hand the company would obviously argue from the earliest stages, on data provided by their own industrial engineers, that fewer men could complete the required processes at a faster rate in order to build their tractors at lower cost and therefore hold or increase their share of the highly competitive market. This meant not only competing with other tractor manufacturers in the UK but also others in Europe including APC's own subsidiaries (especially, in this case, in France). They attempted to convince the shop stewards that, unless their product was competitively priced, a falling market share would inexorably lead to short time working and redundancies whereas a competitive product would lead to more jobs and higher overtime earnings. Eventually, after much bargaining an agreement would be hammered out.

This can best be illustrated by two numerical examples from the negotiation over the operation of the cab tractor track G during the summer and autumn of 1976. The Company, from their own work study tests, wanted the normal Measured Day Work (MDW) rate to be six tractors an hour. The Union argued that four per hour was the most that could be built. One factor in this was the fitting of the radiator. The shop stewards negotiating for Gang 17 claimed that this would take 48 minutes. The Management pointed out that in France it took only 16 minutes; also that gangs 5 and 6F doing similar work on tracks H and Y had an agreed work value of 12 to 18 minutes to fit a radiator. In an attempt to prove their point a scratch team of managers and supervisors without previous experience fitted a number of radiators in times varying from 6 to 20 minutes. To complicate matters, representatives from Gang 5 and 6F declared that the proposals for Gang 17 were nonsense and that they would do the job much quicker if allowed to take over G track. Nevertheless, the Gang 17 shop stewards continued to insist on 48 minutes per radiator and 4 tractors per hour.

In an attempt to drive home the reality of the competition the company organized a visit by a joint management/union delegation to APC in France, where they were producing an almost identical cab tractor more quickly and more cheaply. The British unions, however, argued that this was being achieved by largely immigrant workers who, unlike most immigrant workers in Britain, had no guaranteed right of residence and, being extremely anxious to remain in France, could be exploited as cheap labour. The visit, therefore, merely increased the bitterness and intransigence.

With hindsight, the company's estimates of six tractors an hour was later proved by events to be a fair one but, ironically, the loss of APC(UK)'s share of the market meant that the whole factory, including the gangs which precipitated the dispute, had to go on short time in 1978.

By October 1976 the company had come to the conclusion that negotiations were getting nowhere and that they must bring them to a head. They announced that it would not be worth manufacturing cab tractors for the European market at all unless they could be produced at a competitive price and that this meant six tractors per hour. Gang 17 responded with frequent stoppages, producing only three or four an hour. The company therefore "took them off the clock" which amounted to ceasing production of cab tractors and laying off Gangs 17 and 18 without pay. The unions declared that this was a lockout in breach of procedure and went into dispute just before the Christmas holiday in December 1976. The rest of the Assembly Shop went on strike in support of the two suspended gangs and the entire production of tractors of all types ceased after the Christmas break.

Only the Assembly Shop was on strike but the strikers occupied the factory for six weeks, successfully preventing workers in other shops (e.g. the machine shop) and the staff from working. This was a tactic used by militant shop stewards who knew that the cumbersome union procedure through the District Committee and the Executive Council, each of which met only once every two weeks, would mean many weeks of delay. Rather than wait for this, the stewards called the occupation to halt production while feelings still ran high. After about six weeks the Company applied to the High Court for an interim injunction which was granted and the occupation was ended but by that time the union procedures had been completed to make the strike official and it continued for a further 6 weeks until March 1977.

There was considerable bitterness on both sides. The unions were suspicious that the company had deliberatey engineered the strike because they were having problems with the new tractor and were happy to halt production while they sorted these out; and that they cynically prepared for this by producing more than they needed in the summer and autumn in order to give them six or seven weeks stock to tide them over the strike. (They emphatically deny this, saying that production precisely followed projected schedules.) Much was made of the alleged breach of procedure. The strike leaders probably introduced this deliberately to muddy the issue. They were well aware that the other gangs had little sympathy for the case of Gang 17 on its merits so, to win wider support for the strike they felt that they should make the issue a breach of procedure, on which they could call for union solidarity. This support would be essential to get the dispute nationally recognized by the unions as official.

There was some apprehension of violence. The management went through the picket lines to continue work, but other manual workers who were laid off but who wished to work were paid unemployment benefit, the Social Security Office accepting their plea that they dared not cross the picket line for fear of violence. (This was treated as a test case.) The strike was eventually settled in March 1977. The company agreed that they would not take people off the clock until agreed procedure had been exhausted, and then only after 24 hours notice. They also agreed to refund the holiday pay (about £40) which had been paid before Christmas but which had then been deducted to the great anger of the strikers.

Production was resumed but for the next 2½ years (March 1977–December 1979) G track still produced tractors at only four per hour. Thereafter, many of the strike leaders having left the company and the gang being fed up with low earnings due to the reduced volume of production, it was agreed that production would change to a piecework basis at four per hour.

THE COST OF THE STRIKE

The APC(UK) West Midlands factory lost its entire production for 12 weeks. The planned production was 2000 tractor sets per week (including some 800 built up tractors, the rest being in kits for assembly overseas). The built up tractors were priced at £6500 to £7000 each so the loss of some 10 000 of these alone amounted to £60–70 million in potential sales. Even if the unions were correct in their claim that the company started the strike with six weeks stock of completed tractors this would only have hedged them against half the loss. Against this there were the savings in manual workers' wages, about £5 million, though the company continued to pay the 2000 white collar staff. The cost of materials and inventory and interest costs could also be offset. On the other hand, the company had to bear the cost of holding about 800 cabs, because these had to be accepted under the contract from GKN, thereby tying up £½–£¾ million in capital plus storage costs and the subsequent cost of reworking them after deterioration in store so the net cost to the company will certainly have run into tens of millions of pounds. On top of this was the incalculable cost of diversion of management time to the dispute, both in the run up to the strike and in getting production back into its stride afterwards.

The loss of profit for the company (as distinct from sales) was harder to estimate. The profit margins of home and overseas sales vary widely but probably average a net 10 per cent overall, so a rough estimate of the loss of profit over the 12 weeks would be £6 million.

These, however were only direct costs. The longer term costs were less tangible but very heavy and long-lasting, arising from a reduction in volume of tractor production in UK as a whole and in APC(UK)'s market share. This cannot be blamed wholly on the strike because there was a significant increase in the level of foreign imported tractors and the company realized afterwards that they had made an error of judgement in their marketing, in that the buyer found it cheaper to fit new cabs to existing tractors than buy the newly designed ones with integrated cabs and this was within EEC laws. Due to the recession, moreover, the total (UK) tractor market fell during 1976–81 from 36 000 to 18 000 so APC(UK) sales in UK fell from 12 000 (33 per cent of 36 000) in 1976 to 3600 (20 per cent of 18 000) in 1981 – a drastic drop to less than one third in five years. Taking production as a whole (i.e. including Third World exports, tractor sets for assembly overseas etc.) sales halved, and the West Midlands factory contributed heavily to APC's loss of some £200 million in 1978.

The damage was permanent or at best semi-permanent. With 90 per cent of APC(UK)'s output going for export, the total stoppage of production for 12 weeks sapped the confidence of distributors. European sales (including UK sales) were particularly hard hit because in the crucial months in which the manufacturers were competing to establish their market share for the new cab tractors APC(UK) was out of the race. Moreover, because even when work was resumed the assembly gangs still kept cab-tractor production down to 4 per hour for the next 2½ years, APC found it more profitable to build up the production of the very similar tractor in their factory in France, whose product was soon challenging the APC(UK) product in the UK itself.

The cost to the men on the shop floor in 1977 proved to be greater still in proportion. The 100 men in Gangs 17 and 18 were striking to establish a low production norm which would enable them to earn more when they got onto piecework. At the most optimistic estimate their extra earnings might have been an extra £8 per week or £400 per year.

Their individual wage losses (12 weeks at £80 per week) were about £1000. For men with families this was offset by an average of £17 (maximum £29) Supplementary Benefit, £4 tax rebate and £9 strike pay (withheld until after the strike as otherwise it would have been deducted from their family's supplementary benefit). In other words they were losing some £50 per week or £600 in the 12 weeks. Some of this may later have been made up in overtime but there was never any hope of their doing other than lose money even if they won the strike which, in the short term, they did.

In the long term it was disastrous for them. Of the two gangs which went on strike, the one which trimmed the cabs, Gang 18, has been dissolved since all the cabs are now bought from suppliers in France and Italy. The successors of Gang 17 on Assembly Track G became highly vulnerable, since the sales of cab tractors only required the assembly track to work one week in six. Of the other assembly workers who came out in sympathy with Gangs 17 and 18, many were amongst the 2500 who were made redundant.

On the credit side the JSSC claimed that they won all their points of principle. Their holiday money was repaid; the gangs did resume work in March 1977 at four tractors an hour; and the company agreed not to take them off the clock again without going through full procedures with the convenors and then only after due warning.

Was this a Pyrrhic victory? Was it worth the price which, in a recession already wreaking havoc, helped to cut sales of APC(UK)'s factory to half its productive capacity and to put many more hundreds of men in the dole queues – not to mention those component firms which supply APC(UK)? And was it worth establishing in the minds of APC head office, of distributors and of the clients that APC in France was a more economic and reliable supplier? In the event, Gangs 17 and 18 destroyed themselves and many others.

The tragedy of 1977 was the failure to realize that, rather than higher bonuses, the best interests of both the company and its workforce would be served by higher production at a competitive price. The strike was, once again, an example of one member of the team kicking through its own goal to establish a point of principle.

Should APC management have given way in the first place and accepted the rate of four tractors per hour rather than lose the chances of establishing their share of the new European market for cab tractors? Would it have been worth paying, say, £50 000 in extra bonuses to the two gangs rather than lose millions of pounds in production and profits? Had they cried 'Wolf' too often so that when they did threaten to take action neither the union representatives nor their members believed that they meant it? Did they misjudge the temper of the gangs and their stewards and the solidarity of the other assembly workers – i.e. did they (as the union believed) assume that the strike would cave in in six weeks while they cleared their stockpile? Had they underestimated the 'mobilization of bias'?

For the management, like the union, principles were involved. If they allowed the gangs on the assembly track to get away with deliberately restrictive practices, spending 48 minutes doing a 16 minute job on the radiators and producing 4 tractors instead of 6, when the sole purpose was to establish a false norm to gain higher bonuses and piecework rates, where would it stop? And would it not in the end mean that APC(UK) would lose its share of the market, both to its rivals in the UK and to APC in France? And did they owe it to their work force as a whole to insist on a fair day's work to keep costs competitive and thereby to safeguard both earnings and jobs?

None of these questions, on either side, can be resolved without reverting to the crucial question: how can management, unions and work force be brought to cooperate and accept their common interest in building more tractors more economically? How effective and how frank was managements' communication of the production and marketing realities to its members? Was this communication deliberately frustrated by shop stewards? Did they see in it an erosion of their own power and leadership? And did these stewards deliberately deceive their members over the short and long term losses the strike would cause them in order that they would carry them with them, with the true aim of safeguarding their own and their union's bargaining power? These are the kind of questions to which we shall return in Parts III and IV of this book.

THE GROWING PROBLEM OF REDUNDANCIES

'It's an ill wind . . . '. The strike of 1977 brought home to both sides the price they pay for industrial conflict, particularly in terms of the loss of market share and the consequent redundancies. In their attempts to handle these problems there was a marked improvement in communication and industrial relations, and the production of the next new series of models two years later was introduced smoothly with work values agreed by both sides without dispute. By the time the next conflict arose

4½ years later many of the leading figures concerned had changed. It was thus a largely new team which faced a major dispute over redundancies in 1982.

Up until the end of 1981, voluntary redundancies had been sufficient to meet the necessary cuts but early in 1982 APC(UK) announced that they would need a further 725 redundancies by March, and that some of these might have to be compulsory unless there were enough volunteers. By the middle of March the number of volunteers still fell about 270 short of the number required. The TGWU Convenor summed up the tension as arising from fears on both sides: fear amongst individual workers that there might be a redundancy notice in the next pay packet; and fear by the company that, if they discussed the problem with the Trade Unions at all, they might be "screwed".

On 11 March the unions suggested that the necessary savings should be met by short time working instead of by redundancies and asked the company to set out the costs and working arrangements necessary to implement this. The company replied with a proposal that, instead of 270 compulsory redundancies, 270 people should be laid off on the last-in-first-out(LIFO) principal. Meanwhile anyone accepted for voluntary redundancy would be replaced by reducing the number laid off. At the end of the layoff period, and remaining shortfall would be made up by compulsory redundancy. Those laid off would receive their statutory guaranteed minimum of 32 hour wage (i.e. 80 per cent of the full wage). This would amount to an average wage loss of about £25 per week for the 270 laid of. The company was also willing for the lay off costs to be spread over the factory as a whole which would require a deduction of £12.52 per person per week.

The union objected to the LIFO principle and to open-ended layoffs – and to any compulsory redundancies at all, so the company offered an alternative 'rotational' scheme whereby people would be laid off for one month in turn, those for the first month only being selected on the LIFO principle. The unions rejected those proposals and there was a walk-out on 22 March. Hope of a settlement revived and the union agreed to a return to work and resumed negotiations on 24 March. The unions proposed that the costs should be met by the whole work force going onto a 4-day week, still receiving about 90 per cent of normal pay, which would mean that each man would forfeit the same average £12 to £12.50 per week. The company argued that, while they could meet their production commitments if there were 270 lay offs spread over departments, they would be unable to meet them if everyone worked only a four day week: also that this would not in fact make the savings which were required to make for a number of reasons, partly because they would be paying 90 per cent wages for 80 per cent output (4 day week), and not least because the fringe costs (pension contributions etc) would still have to be paid and these added some 20 per cent on top of a person's pay. Protracted negotiations on 30 March were adjourned until next morning (31 March) when they were resumed but broke up without agreement.

Just before the negotiations broke up on 31 March the management handed out compulsory redundancy notices to 170 people (the number required having been reduced by ongoing voluntary redundancies). The union negotiators regarded this as a breach of faith as they were still sitting round the table. The company, however, argued that it was quite clear by then that no agreement would be reached, that as soon as the negotiators left the conference room they would call an immediate walk out and that it would then be virtually impossible to find the people to whom to issue the redundancy notices. The union leaders did indeed call an immediate walkout and the 2½ week strike began in an atmosphere of great bitterness.

During these 2½ weeks the company obtained a High Court injunction to evict the strikers, the Union succeeded in getting a stay of execution and the company appealed direct to the work force by commissioning a secret ballot to be carried out by the Electoral Reform Society. At the same time they issued an ultimatum that unless the strike was called off by 5 April they would withdraw all the voluntary redundancies.

All of these developments, however were overtaken by a steady increase in the number of volunteers for redundancy and – in the latter stages – by what appeared to be a growing drift back to work. The trade union itself cooperated in the search for voluntary redundancies and, though the premises seized and picketed did include the management offices, the pickets allowed the personnel management staff to have access to the men occupying the factory to try to persuade more people to volunteer. By the middle of April the shortfall had been made up by voluntary redundancies and the balance of savings was mostly met by the union accepting more flexibility over skills. The factory resumed work on Monday 19 April

What did the strike achieve? The union, by playing for time, managed to avoid compulsory redundancies. The short term price to the work force was 2½ weeks wages – probably of the order of £300 per man which, though cushioned by pay-in-hand etc (see Chapter 7) would ultimately be at the expense of their family budgets.

The indirect costs were probably far greater. The loss of production was worth £10 million. The spring is the time when the demand for agricultural tractors is highest so many of these sales will have been irrevocably lost – mainly to factories outside the UK. Worse still APC(UK) will have forfeited future orders because customers are always wary of buying from manufacturers whose deliveries are made uncertain by industrial disputes. This has been a major cause in the decline of the market share at BL in the 1970s, and clearly also a factor for APC (UK) since the 1977 strike.

Equally sad was the damage to the cooperative industrial relations atmosphere patiently rebuilt during the 4½ years since 1977.

Why was it that, when agreement seemed so close, both sides paid this price rather than bridge the gap? Inevitably their interpretations in retrospect are very different. On the union side the probable explanation is that they believed that the company would still make some people compulsorily redundant at the end of their layoff period, especially if they had been selected on the LIFO principle; and that these people, being the youngest and least qualified, would be cheapest to pay off and worst equipped to find other employment. In this respect it is significant that the majority of voluntary redundancies were amongst older people, who were able to claim very attractive arrangements for part conversion of their pensions and often went out with a handout of about £5000 in their pockets. Some £1500 of this could be claimed back by the company from the government, so the cost to them might be some £3500. Since the average cost of each employee including fringe costs was about £8500 the company could expect to save their £3500 in less than six months.

The company, on the other hand, were genuine in their belief that they could not meet their production targets with 3000 men on a 4-day week – i.e. at 80 per cent of the factory's output at 90 per cent of the wage bill. Though the wage cost would be the same as laying off 270, the second alternative would enable the factory to keep going at not far short of full production for 5 days a week albeit with a reduced staff. Moreover the 4-day week would not meet the savings required because, amongst other things, of the continuation of fringe costs as explained above. The company also probably appreciated that these were not necessarily the last redundancies there would be – and indeed they did have to announce another 400 in November 1982. In the event, these 400 were cut slightly due to an upturn in demand, and the requirement was again met entirely by voluntary redundancies by the summer of 1983. It is to the credit of all concerned that there have not in fact been any compulsory redundancies at all in APC (UK).

The hard-pressed personnel and industrial relations staff and convenors have meanwhile been doing their best to rebuild their damaged relationship yet again. The convenors are convinced that their opposite numbers are often anxious to accommodate to their demands but that they are prevented from doing so by higher management or the corporate headquarters, though the management deny this and accept responsibility for their own decisions. Personal relationships at working level remain friendly and, in addition to a modest pick up in worldwide demand, APC (UK) have also clawed back a small part of their lost market share from their competitors.

Nevertheless, if they could have found a way of avoiding kicking two expensive own goals in 1977 and 1982 this market share would no doubt be greater because some of the tractors now manufactured overseas (whether by rival corporations or by other APC subsidiaries, eg in France) would be being manufactured in UK. So, in the end, these own goals have inevitably been paid for out of family budgets and by more British workers joining the dole queue.

10 The Motor Industry

EVEN WORSE THAN THE WORST CASE

In 1975, the year British Leyland was nationalized, the Central Policy Review Staff (CPRS) published a report on the future of the British car industry.[1] It catalogued its decline, compared its performance with that of its rivals and analysed the reasons. It proposed remedies, gave dire warnings of the results of failure to rectify the faults and predicted a 'best case' and a 'worst case' for 1980 and 1985. In the event, the actual performance in 1980 was even worse – indeed far worse – than the worst case CPRS had predicted (see Table 10.1).

TABLE 10.1 *Projected and actual performance of the British motor industry*[2]

			1974	1980			1985	
				Forecast			Forecast	
				Best case	Worst case	Actual	Best case	Worst case
	Sales in Britain							
1	New registrations	(millions)	1.27	1.65	1.55	1.51	1.90	1.80
2	Import share	(%)	27.90	28.00	45.00	56.70	28.00	45.00
3	British-built sales	(millions)	0.92	1.19	0.85	0.65	1.37	0.99
	Exports to EEC/EFTA							
4	New registrations	(millions)	6.44	8.45	7.25	7.76	9.70	8.60
5	British share	(%)	3.20	4.00	3.00	1.80	4.50	2.80
6	British-built sales	(millions)	0.21	0.34	0.22	0.14	0.44	0.24
	Exports to Rest of World							
7	To North America	(millions)	0.08	0.10	0.08	0.03	0.14	0.08
8	To rest of world	(millions)	0.27	0.19	0.14	0.18	0.19	0.14
	Total British Car Production							
9	3 + 6 + 7 + 8	(millions)	1.48	1.82	1.29	0.99	2.14	1.45

The total British production in 1980 (serial 9, 0.99 million) was little more than half what the CPRS had stated as the minimum break even production (1.80 million) and well below the worst case forecast (1.29 million). The share of the home market taken by imported foreign cars (serial 2, 56.70 per cent) was not only more than the worst case (45 per cent) but in real terms worse still since a growing number of 'British' cars were being assembled from largely imported components, accounting for 67 000 of the 990 000 total British production (serial 9).[3]

The numbers employed in the major UK motor companies declined, predictably, but so too did the position of car workers in the national earnings league where they had traditionally been amongst the highest. A work force of 304 000 in 1971 fell to 196 000 in 1981, and in the same period, the earnings of male manual car workers compared with the average in all industries and services (100 per cent) fell from 126 per cent to 99 per cent – below average manual earnings for the first time.[4]

There were many reasons for this decline. The British motor industry had achieved the highest output per man in Europe in 1955 but it was overtaken by all its competitors in turn (see Table 10.2).

TABLE 10.2 *Vehicles per employee per year*[5]

Country	*1955*	*1965*	*1973*
UK	4.2	5.8	5.1
USA	11.1	13.9	14.9
Germany	3.9	7.1	7.3
France	3.6	6.1	6.8
Italy	3.0	7.4	6.8
Japan	1.2	4.4	12.2

Productivity was particularly bad on the assembly lines where, even building identical models using identical capital equipment, output per man in 1975 was little more than half that in Continental Europe.[6] British plants used 50–70 per cent more maintenance men yet lost twice as many production hours due to breakdown.[7]

Though there were in Britain far too many plants and far too many models, the greatest cause of poor productivity was lack of motivation arising from poor industrial relations, and the consequent disruption of production by strikes. In 1970–78, 20 per cent of all days lost in strikes in all industries and services were in the motor industry, though it only employed about 1 per cent of the labour force. Even the mining industry, with a comparable labour force, and despite two national strikes in 1972 and 1974, accounted for only about half as many (11.5 per cent).[8]

Internal and external labour disputes caused an average 75 per cent of the lost production in 1970–74 in the UK motor industry, far more than mechanical breakdowns (about 5 per cent) and production inefficiences (about 15 per cent).[9] This was not due to lack of effort by its British managers, who spent almost 50 per cent of their time dealing with labour disputes compared with 5–10 per cent quoted by motor industry plant managers in Belgium and West Germany.[10] There was, however, a long history of lack of trust and communication between management and labour force and a strong resistance by shop stewards to any such communication other than through them. (This blockage has largely been rectified in the 1980s as is discussed later in the chapter).

An American study has shown that repetitive work in which the pace is dictated by the machine is the most boring and frustrating of all types of employment, closely followed by repetitive work in which the operator dictates the pace.[11] The cure for this must lie in the inhuman and degrading slavery of the assembly line being replaced by robotics – and in finding more worthwhile employment for the people released from it, as is discussed in the final chapter of this book.

In the period 1970–78, the British motor industry was in the grip of a vicious circle. Bad labour relations led to low productivity, low profit, low investment, continued frustration and thence to worse labour relations. The first major break in this vicious circle came with the arrival of Sir Michael Edwardes at British Leyland in 1977.

BL – THE CRISIS 1977–9

The British Leyland Motor Company (BLMC) was formed in 1968 by the amalgamation of the successful Leyland Motor Corporation, making mainly trucks and buses, and the British Motor Corporation (BMC) which was itself an amalgam of almost every major car manufacturer in Britain. All the famous makes, including Leyland, Scammell, Albion, AEC, Guy, Coventry Climax, Daimler, Jaguar, Lanchester, Alvis, Meadows, Rover, Standard, Triumph, Riley, Austin and Morris were incorporated into one giant corporation, largely on the initiative of the Prime Minister Harold Wilson, Tony Benn and Sir Donald Stokes, then

Chairman of Leyland.[12] The idea was to achieve the economies of scale on the lines of the giant American motor corporations (though they too have had more than their share of troubles) so that engines and other assemblies could be interchanged. BLMC, however, never really succeeded in this and suffered all the disadvantages and few of the advantages of size.

By 1974, BLMC was in deep financial trouble. It was baled out by what amounted to nationalization in 1975 when, following an inquiry and the formulation of a plan by Lord Ryder, it was taken over by the National Enterprise Board (NEB). At this time it embraced 34 plants, and 58 hourly-paid bargaining units, and there were wide disparities in wage structure, conditions and benefits.[13] There were 17 different trade unions directly involved – as many as in the whole of German industry.

BL Cars' share of the UK market continued to fall, from 32.7 per cent in 1974 to 24.3 per cent in 1977 and *The Economist* reported that 'Communications with the shop floor and outside union officials remain as awful as ever. . . . In six out of the last seven months (BL) has been catastrophically hit by disputes.'[14] It was in this climate that Sir Michael Edwardes was appointed Chairman in November 1977.

Edwardes was 48 years old, a small, dynamic South African who had a record of success as Chairman of the Chloride Group. After a few weeks assessing the situation, he made a plan which involved the closure of 7 plants, an initial cut of 12 500 in the workforce (he was to cut 82 000 in all during his 5 years as Chairman of BL), and decentralization into more manageable product groups. He radically reorganized the structure of managers, disposing of many who were unable or unwilling to manage in accordance with his philosophy or who had been unduly demoralized by the years of failure in BL. Within a few days of arrival he introduced scientific psychological written tests, for his top 300 managers' posts. During the first few months he recruited 60 new ones and moved 150 of the remaining 240 into different jobs. Most of the others resigned or were asked to leave with termination payments. It was clear to all that he was going to be as tough with management as with anyone else and this did not pass unnoticed on the shop floor.

He decided at an early stage to explain BL's situation directly to a delegate conference of union officials and shop stewards. This was held at Kenilworth on 1 February 1978 and 720 attended. He explained his plan, warned of the coming closures and the initial manning reduction of 12 500 and persuaded them that, unless this were done the Government, which had poured hundreds of millions of pounds into BL since 1975, would write it off and there would be no jobs at all. He announced at the start of his speech that it would be followed by a debate and then a vote of confidence. Though two militant shop stewards spoke against the plan, others confirmed that their own (Labour) MPs had convinced them that Parliament was indeed losing patience. The delegates passed a vote of confidence in Edwardes by 715 votes to 5, and this received massive publicity.

Two days later there was a conference of 2000 BL dealers at Wembley. Their morale, too, needed restoring after years of frustration – delays in deliveries caused by industrial disruption and decline of customers' confidence. The shop stewards massive vote to endorse the new strategy provided the required reassurance.

The first real crunch, however, came within a few days of the Kenilworth and Wembley conferences with the closure of the 103 acre assembly plant at Speke, near Liverpool. In addition to being too far from the main components factories in the West Midlands, Speke had an appalling industrial relations record and was in the eighth week of an unofficial stoppage at the time of the Kenilworth meeting. When the closure was announced many people were sceptical. There was a widespread belief that a closure in an area of high unemployment would be politically unacceptable and that more government money would somehow be found to bail BL out. The closure, however, went ahead and the workforce, realizing that further rearguard actions would merely lose them more money to no effect, voted to accept the closure terms.

Edwardes considered it vital to establish BL's credibility in this respect, not least to support the moderate union officials and stewards who were ready to provide leadership in reducing disruption and restoring profitability. The watershed came a few months later at the truck plant at Bathgate in Scotland. The company had spent £22 million in installing modern plant. The men came out on unofficial strike to demand extra money for working the new machinery. The moderate union leadership refused to support the strike but one of the shop stewards was reported in the press as saying 'Stay out, lads. They've given way before and they'll give way again.' Edwardes went on television and announced that the losses during the six week strike had cost £30 million and added:

> If we say we will not meet the demands of the strikers because we cannot, and should not, we mean it. Furthermore, I tell you now that investment at Bathgate will be reduced by the amount of cash flow we have lost due to the strike. We simply cannot pay twice. I mean what I say. We will not cry wolf.[15]

Five days later, with the strike still in progress, the shop stewards were informed that a further £32 million of investment due at Bathgate would not go ahead and would be diverted to other parts of the business. The strikers thereupon returned to work with no concessions whatever. Although the dispute had cost both the strikers and BL (i.e. the taxpayers) a lot, Michael Edwardes commented: 'One important achievement had been gained: people were beginning to believe us.'[16]

Though not as bad as 1977, the flow of production in the first six months of 1978 had been disrupted by 346 disputes. The most serious damage was being done by a tiny hardcore of disrupters. Out of 13 major interruptions during this period, 12 had involved a total of only 160 out of a workforce of 190 000. This seemed to confirm the impression that the leaders of this hard core were either politically motivated or endeavouring to extend their own power by dominating the management. On 1 September 1978 Edwardes publicly launched an attack on what he described as 'The militant extremist minority within BL who were putting the company's recovery in real danger'[17] and began a process which was to reach its crisis a year later.

BL – THE BEGINNING OF RECOVERY

James Callaghan and his industry secretary Eric Varley gave unstinting support to Edwardes in his fight to make BL a viable concern and continued to inject money to cover his reorganization period. They also supported his stand against deliberate attempts to make it fail. In May 1979, however, Mr Callaghan's Government fell and Mrs Thatcher was elected. While Edwardes could certainly rely on continued support against militancy, however, he could not be sure that the Conservative Government would give the same financial support since this was against their philosophy of not propping up lame ducks. Moreover, their return to power increased the value of the pound and this had a disastrous effect both on BL's exports and in making the British market more attractive for foreign manufacturers. Edwardes therefore had to ask the new Government for the injection of a further £300 million from public funds; he argued that the alternative would be the closure of BL; and that the redundancy payments for BL workers and others made redundant in associated industries and in the regional infrastructures would cost about £1000 million in the first year.[18]

The Government agreed to provide the £300 million as part of a total of £1000 million over the next few years on the basis of a Recovery

Plan worked out in the summer of 1979.[19] This plan was designed to enable BL to break even by 1983, but at the expense of closing 13 plants and losing a further 25 000 jobs during the next 2½ years. It was the publication of this plan which brought about the most decisive confrontation in Edwardes' five years as Chairman.

Foremost amongst the trade union militants was Derek Robinson. He was the AUEW Convenor at BL's main volume car plant at Longbridge, and a member of the Communist Party of Great Britain. It was estimated that during Robinson's 30-months as Longbridge convenor there had been 523 disputes, with the loss of 62 000 cars and 113 000 engines, worth £200 million.[20]

The Recovery Plan was published on 10 September 1979 and was at once rejected by the unions. BL responded by putting the issue direct to the workforce, to be 'decided by 150 000 employees, not 250 shop stewards'. The Electoral Reform Society was commissioned to conduct a secret ballot on the question 'Do you give your support to BL's Recovery Plan?' The result was announced on 1 November 1979: 80 per cent of employees voted, 87.2 per cent of them (over 106 000) voting 'Yes' – a vote of 7 to 1 in favour.

A few days later BL management received, anonymously through the post, a copy of the minutes of what purported to be a meeting held shortly after the Recovery Plan had been announced, between Communist Party officials and BL shop stewards, though the Minutes did not list Robinson as being one of those present. They outlined a strategy for thwarting the Recovery Plan and, though there was no proof that the Minutes were genuine, the strategy and the wording seemed to bear a striking resemblance to those in a pamphlet distributed in November and signed by four shop stewards including Robinson. This called for active resistance to the plan, including refusal to accept transfer of work from one plant to another:

> This does not mean a passive role by the receiving plant. They must be actively involved. In other industries, like UCS, work-ins and occupations have been necessary to prevent closure. If necessary we shall have to do the same.

BL issued a statement in reply on 19 November stating:

> By publishing such a booklet calling for disruptive action the people concerned are deliberately undermining the company's recovery programme, threatening both the market share and confidence in the company's future.[21]

Since this programme had been accepted by a 7 to 1 vote by the workers whom Robinson was supposed to represent, the Longbridge plant manager invited him to withdraw his name from the pamphlet but he refused, making it clear that, ballot or no ballot, he opposed the plan. Like most convenors, Robinson was a paid employee of the company but free to work full time on his trade union duties. The company, convinced that his active opposition to the Recovery Plan no longer represented the views of his members and was imcompatible with his continued employment by the company, immediately dismissed him.

Commentators almost to a man expressed astonishment and suggested that Edwardes had played into Robinson's hands by making a martyr of him. First reactions seemed to support this view. There was an immediate unofficial strike led by Robinson's fellow shop stewards at Longbridge and workers at some other factories came out in sympathy. Although about 1000 of the 20 000 Longbridge workers pushed their way through the picket line with the purpose of working it was clear that BL was in for another strike and that it was likely to be made official. The AUEW called for Robinson's reinstatement but Edwardes would not agree to this, and showed the union officials a draft letter he had prepared, to be sent to all employees, warning them that if they did not return to work on 4 December they would be deemed to have broken their contracts and would be dismissed. The union suggested a compromise that Robinson be kept on the payroll pending a hearing but BL would not accept this and it was eventually agreed that the company would give him a weekly *ex gratia* payment equivalent to his normal wage while the AUEW conducted their own inquiry; and that the men would meanwhile return to work.

The AUEW Inquiry completed its report in the first week of February. It was highly critical of Robinson but called on the company to reinstate him on the grounds that they had not followed the correct disciplinary procedure. BL announced that his behaviour had made this impossible and that a strike would not change that decision.

A full scale strike seemed inevitable but was delayed by a spontaneous petition by 2000 Longbridge workers on 8 February calling on the AUEW to sound out the views of their members before calling the strike. To complicate matters, the normal annual 1979–80 pay talks were continuing quite separately at the time and broke down on 15 February, after the workforce had rejected the company's offer by 3 to 2 in a secret ballot. Almost every commentator predicted that, when the union called its mass meeting to vote on whether to strike over the reinstatement of Derek Robinson there would be an overwhelming majority for strike action.

In the event, the mass meeting was held on 20 February and voted overwhelmingly against striking over reinstatement. The vote was estimated at 14 000 to 600.[22] There can hardly ever have been a more dramatic example of the rejection of the lead of militant shop stewards by their members.

This was a major victory for Edwardes. Though BL lost a further £531m in 1981,[23] the company was clearly on the road to recovery. Industrial disputes became rare. Two new models, the Metro in 1980 and the Maestro in 1983 were successfully launched with largely robotic assembly lines. The robotics and the new working practices were accepted, though there were still some serious stoppages. One of the worst of these was a prolonged strike over whether or not the men at Cowley were to be allowed to start work three minutes late and stop three minutes early to allow time for 'washing up' – which was a practice unique to Cowley and not applicable to other plants. This dispute halted production on 27 March 1983 for 31 days, with the loss of production of 19 000 cars (mainly the new Maestros) valued at £100 m. Stocks of completed cars were large enough to avoid much direct loss of sales but the strike did set back the revival of confidence in BL at home and overseas. The factory returned to work on 27 April with a formula for resuming talks but the union leaders were dissatisfied with the company's eventual offer. The stewards called for a resumption of the strike but their call was rejected on 24 June in a secret ballot by the work force, each of whom had lost some £500 during the strike from which they gained virtually nothing.

Generally, however, this was only a brief aberration after a rapid improvement in productivity. In 1980 the Longbridge factory produced 7 cars per man; in 1981 this rose to 17 and in 1982 to 25. The proportion of sales in Britain taken by foreign built cars fell slightly in the first half of 1983 compared with 1982 despite an increase of 18 per cent in new car registrations and in the same period BL's sales volume increased more than this, by 20 per cent.[24]

THE WAY AHEAD IN THE MOTOR INDUSTRY

It was not only in the volume car and commercial production groups that Sir Michael Edwardes revolutionized BL, but also in the up-market sports car sector which accounted for the bulk of exports to the USA, especially Jaguar.

Jaguar had been losing money with declining sales for some years, not least because their cars had acquired a deplorable reputation for unreliability even amongst their most loyal enthusiasts. Early in 1980 Edwardes persuaded John Egan, who had left the company some years earlier, to rejoin Jaguar as managing director. Egan at once tackled the problem of communication with the shop floor and quickly had the shop stewards working with him. They all realized that, unless they could produce a reliable product more economically Jaguar would disappear. To help achieve this, 50 quality circles were set up.

The decline was dramatically reversed. In 1976, Jaguar's record year, 31 000 cars had been produced. By 1978 this had fallen to 26 500.[25] In 1980 a workforce of 10 000 produced only 14 000 cars. By 1982 the workforce had been cut to 7200 but they produced 22 000 cars – i.e. an increase of well over 100 per cent in productivity in 2 years. At the same time the reliability of the product improved and export to the USA increased by 55 per cent in 1981 over the 1980 level, and by 100 per cent (i.e. doubled) in 1982.[26] In the first six months of 1983, worldwide sales had risen by 42 per cent over the first six months of 1982 and sales to the USA (7733 cars) by 73 per cent.[27]

The most impressive lesson for British industry is that, because their increase in productivity and reliability produced a more marketable car, they were able, after reducing their staff to start taking on more labour in 1983 in order to meet the increased demand. This must surely indicate the way in which Britain should approach the problem of unemployment.

Meanwhile the process begun by BL as a whole in 1980–83 must continue, in particular the robotization of production lines which has been so successful in launching the Metro and the Maestro. This will gradually phase out the degrading misuse of manpower on repetitive assembly line work. It is the boredom and frustration of this, especially on tracks where noise is high, which has produced the climate in which the motor workers have responded to the leadership of militant shop stewards intent on maintaining conflict as a means of maintaining their own power. Thanks largely to the efficient commercial management and strong leadership of Sir Michael Edwardes, the response to militancy has greatly declined. Repeatedly, the rank and file have rejected their shop stewards' calls to strike. Whether they will continue to do so, as unemployment declines, remains to be seen. The phasing in of robotics should reduce the risk.

On a broader point, Sir Michael Edwardes has estimated that, in the past 20 years, Britain's share of the world market in goods as a whole has halved and that each 1 per cent lost has meant another 250 000 jobs lost in Britain. Similarly, each 1 per cent of Britain's home market regained from foreign imports for British manufacturers means another 80 000 jobs. He is convinced that we can win back both 2 per cent of overseas markets and 2 per cent of the UK market, which would provide another 660 000 jobs.[28] That such is not inconceivable has been proved by BL in general and Jaguar in particular.

The start of 1984 produced evidence that the British motor industry was once again becoming competitive. The total British market for cars in 1983 was 1 780 000, up 15 per cent on 1982. Austin Rover led the way, with sales 20 per cent up on 1982; they produced 450 000 cars, which was 44 per cent of all the cars produced in Britain, taking 18 per cent of the market. Ford took 29 per cent but over half of these were imported from Ford factories in Germany, Belgium and Spain.

Jaguar, due for privatization in 1984, produced even more impressive figures. World sales were 29 100, up 33 per cent on 1982, at a showroom value of £550 million. Exports to USA were 15 815, up 53 per cent on 1982, and to Germany up 44 per cent. John Egan estimated that to keep pace with Mercedes-Benz in automation, they would need to invest at least £40 m a year for five years and, with estimated profits of £30–60 m in 1983, the prospects were promising.

The British motor industry's fight for survival will continue; it will only be won if investment is at least equal to that of its rivals, and its unit costs remain competitive. Landrover had a bad year, with production falling from 52 000 in 1982 to 41 000 in 1983, largely due to Japanese competition in Third World countries. Many years with a virtually unchallenged reputation probably made the company complacent so that investment was inadequate and Japanese equivalents in some cases cost only half as much so, while Landrover's share of the overseas market fell to 25 per cent, the Japanese raised their share from 25 per cent to 50 per cent in four years. Sadly, the reaction of the workforce was to start 1984 with a vote to strike over a pay claim. Nevertheless, plans for investment may enable Landrover to recover its position.

The most encouraging achievement of all in 1983 was that the Austin Rover workforce raised their productivity from 7 cars per man in 1979, through 25 in 1982 to 40 cars per man in 1983, equal to the highest in Europe.[29] Therein lies the only road to salvation.

11 Public Services

A STRIKE-PRONE SECTOR

The majority of the most serious strikes in Britain in the past 20 years have occurred in the publicly owned sector, i.e. public services and nationalized industries. This would astonish the Victorians because the right to strike and the immunities for pickets were pioneered by Britain in the 19th Century in the interests of public order. Alarmed by disorder and revolution on the Continent, the British Parliament, encouraged by the more enlightened industrialists, aimed to give British workers a legal means (in place of violence) of balancing the excessive power of private employers by hitting their profits. In the late 20th Century this pressure has largely been switched onto the public, both in terms of deprivation of services and picking up the bill as taxpayers.

A comprehensive Department of Employment study of strikes in Britain bears this out.[1] The average number of working days per 1000 employees lost each year for 1970–75 was by far the highest in coalmining (9200) and docks (4200) followed by motor and tractor manufacturing manufacturing (3500 and 3200) while the median for all British industries was only 300, and in the average year 98 per cent of manufacturing firms had no strikes at all.[2] Five industries (coal, docks, car manufacture, ship building and iron and steel) accounted for a third of the days lost though they contained less than six per cent of those employed.[3]

The effect of public service strikes is compounded by an unmeasurable gearing factor in that other industries and individuals are hampered in their own work by denial of fuel, transport and materials. Also, since so many people are affected, such strikes can be guaranteed to attract publicity.

It is probably for these reasons that the public services and nationalized industries (including BL cars, nationalized since 1975) have attracted more than their share of politically motivated strike organizers. It is no coincidence that many of the best-known militant trade unionists (e.g. Arthur Scargill and Derek Robinson) have been in these industries. This was further illustrated in August 1983 when a well organized infiltration of BL by members of a Trotskyist organization (several of whom concealed their university degrees and produced fake references) was uncovered.[4]

Britain has far less restrictions on public service strikes than most other industrial countries (see Chapter 5). Such restrictions as do exist are designed to deal with threats to life and public health rather than to safeguard employment or economic activity. Thus it is a criminal offence for strikers to cut off gas, water and electric power,[5] though they can reduce supplies by 'working to rule' as the Power Station Workers did in 1970. Under the Emergency Powers Act of 1920 as amended in 1964, the Government can declare a State of Emergency

> If at any time it appears . . . that there have occurred, or are about to occur, events of such a nature as to be calculated, by interfering with the supply and distribution of food, water, fuel or light, or with the means of locomotion, to deprive the community, or any substantial portion of the community, of the essentials of life, Her Majesty may, by proclamation . . . declare that a State of Emergency exists.

The principle effect of such a State of Emergency is not, in practice, to make the strike a criminal act, but to enable the Government to requisition any of the equipment (e.g. fire engines or ambulances) and use other people to operate it (usually the armed forces); also to order power cuts or reduced speed limits on the roads to save energy.

In addition to electricity, gas and water workers, certain other workers have restrictions on striking (e.g. merchant seamen at sea) and others are banned from striking altogether (e.g. the armed forces and the police). Britain is one of the few countries in the world in which firemen can legally strike.

THE RIGHT TO STRIKE

In 1980 Dr L. J. Macfarlane[6] pointed out the difference between strikes to demonstrate opposition to government policies (usually one-day protest strikes) and strikes aimed to coerce a government into changing its policies. He further subdivides coercive strikes into 'economic' strikes in furtherance of trade union interests and objectives, or 'political' strikes in pursuit of wider objectives. He analyses the 1926 General Strike as a major "economic" strike and concludes that,

> Even if the TUC's proposals were reasonable, and the Government's response open to severe criticism, that did not mean that the trade unions were entitled to bring the country to a halt in order to force the Government to adopt policies to which it was opposed. In terms of democratic theory, as distinct from Marxist Theory, the case of the 1926 General Strike is at best a doubtful one.

Since then, the proportion of the population employed by the Government has greatly increased and it has become by far the largest single employer. Macfarlane considers it unjustifiable for coercive economic strikes to be conducted in such a fashion as 'to cause deep and serious harm to vulnerable sections of the population'.[7]

He regards coercive political strikes as justifiable against autocratic or self-appointed governments but not against democratically elected ones. He cites the example of the Kapp Putsch in Germany in 1920, in which a right wing politician supported by a private army seized control of Berlin in an attempt to oust the elected Government, and was defeated by a general strike; and notes that Hitler might himself have been ousted by a general strike had he not decreed severe penalties for anyone who provoked such a strike. More recently, most of the people of the world would justify the actions of Solidarity in their attempt to force the Polish Government to recognize independent trade unions and to tolerate at least some legal means of opposing the Government of the Polish Communist Party.

Macfarlane, however, does not extend this justification to democratic societies:

> The unpalatable truth that needs to be asserted is that coercive industrial action against the Government in furtherance of directly political objectives is a challenge to the democratic political system. It is dangerous precisely because it constitutes a direct challenge to the right of an elected government to give effect to its policies, even when there is no question of such policies involving interference with basic union rights or interests, the fundamental rights of citizens or the maintenance of the democratic system itself.[8]

BRINGING DOWN GOVERNMENTS, 1974 AND 1979

The 1970s saw the high noon of trade union power and the overthrowing of two governments, one Conservative and one Labour, primarily as a result of public service strikes.[9]

Sir Denis Barnes, formerly Permanent Secretary of the Department of Employment, commenting in the immediate aftermath of the fall of the Callaghan Government in 1979, predicted that the unions would attempt to increase their power still further, resisting any 'permanent incomes policy, either voluntary or statutory, involving permanent restraint on the use of their industrial power' and that,

> Thirdly, they will pursue their syndicalist objectives and aim to increase their industrial power in the public sector through the extension of public ownership and changes in the management of public enterprise. . . . Fourthly they will aim to increase their political power; and in particular to have more effective influence on (control, if possible, on some issues) Labour Governments.[10]

He also predicted the lines on which the newly elected Conservative Government would attempt to restrain this growth of union power without repeating the conflicts of 1972–4.

The attempts by the Heath, Wilson and Callaghan Governments to control wage inflation and trade union power were discussed in Chapter 3. That all three were defeated was almost entirely due to the trade unions' use of public service strikes.

The Heath Government were particularly unsuccessful in handling their public service strikes. The first challenges came in 1970 within a few weeks of their election, by dockers, council workers and power station workers. All were referred to Courts of Inquiry which in each case recommended increases far in excess of the Government's intended norm of an annual rise of 8 per cent. The Power Station workers staged a work to rule in December 1970 in pursuit of their demand for 25 per cent. A State of Emergency was declared and, to keep industry going, domestic consumers were subjected to several long winter evenings each week without electric light or heat. The public opinion polls recorded new records in unpopularity of the unions but Lord Wilberforce's Court of Inquiry made an award which *The Economist* estimated as worth 18–19 per cent.[11] The Government's only success was against the Union of Post Office Workers (UPW) in the early months of 1971. This in fact, rebounded in one sense because the postmen, being in daily and personal contact with their neighbours, and led by a moderate and popular leader (Tom Jackson) were reluctant to make the public suffer so their failure had the effect of discrediting moderation in the eyes of the unions.

The Industrial Relations Act of 1971 made matters worse. The Government tried the compulsory secret ballot against the railwaymen but, as recorded in Chapter 3, the reaction was a closing of ranks in the union. Imprisoning shop stewards in the docks caused such a reaction that the government were obliged again to climb down, so in 1972 the miners were confident of victory when they began the first of the two strikes which brought down the Heath Government.

The 1972 and 1974 miners strikes both began as 'coercive economic' strikes and indeed achieved their economic objects. The Government's norm of 8 per cent was irrevocably shattered in 1972 by a settlement of 27 per cent after the first violent mass picket at the Saltley Coke Depot. The 1974 strike continued throughout the General Election which it precipitated and was settled by the incoming Wilson Government with increases of 22 per cent (skilled) to 32 per cent (unskilled) leading to an increase of 48 per cent in coal prices.[12]

The real significance of these two strikes was that they did develop into 'coercive political' strikes. This was made clear by Arthur Scargill, at that time organizer of the flying pickets at Saltley and, by 1974, President of the Yorkshire Miners. In 1975, in an interview, he was recorded as saying,

> The biggest mistake we could make is that of suggesting that a wage battle is not a political battle. Of course it is. . . . You see we took the view that we were in a class war. . . . We were fighting a government. Anyone who thinks otherwise was living in cloud cuckoo land. We had to declare war on them and the only way you could do that was to attack the vulnerable points. They were the points of energy; the power stations, the coke depots, the coal depots, the points of supply. And this is what we did. . . . The miners' union was not opposed to the distribution of coal. We were only opposed to the distribution of coal to industry because we wished to paralyse the nations economy. Its as simple as that.[13]

The miners were lucky in that their strike coincided with the 1973–74 Arab Israeli war, in which the Arabs cut oil supplies and quadrupled the oil price on 1st January 1974. North Sea Oil was not yet on stream, nor had the Government the cushion of 3 months oil stocks which the

EEC subsequently introduced to counter future use of the Arab oil weapon. Coal stocks at power stations were inadequate to last through the winter. Once again, a State of Emergency was declared and industry was restricted to a 3-day working week. (This provided an interesting side effect, with deep implications for British industry because, though working hours fell by 40 per cent output fell by only 25 per cent in January and was climbing again by February, some firms reaching 100 per cent again before the 3-day week was over.[14])

If the miners did not succeed in 'paralysing the nation's economy' they did succeed in bringing down the Government. Edward Heath called a General Election for 28 February 1974 on the issue of 'who runs Britain? Government or Unions?' Initial public opinion polls showed a big lead for the Conservatives but this rapidly dwindled and, though the Conservatives got more votes than Labour, they won four fewer seats and Wilson became Prime Minister. The reason for this loss of support was that many voters, while clearly resenting the wielding of political power by the unions, had a real feeling of disquiet that the Heath Government was unable to find the answer to union power and that the country was on the verge of economic and social collapse. Wilson managed to convince enough people that he would make a 'social contract' with the unions which would work.

In general terms he (and Callaghan who took over in 1976) did make it work, but only for a short time in 1976–77. Fuelled by the 22–32 per cent settlement for the miners, inflation climbed from 12 per cent (annual rate) at the end of 1973 to 27 per cent in the third quarter of 1975, unemployment almost doubling over the same period.[15] Facing the prospect of galloping inflation and a potential financial collapse like the one which destroyed the Labour Government in 1931, the TUC carried its member unions with it in agreeing to Wilson's social contract, which proved highly successful in the form of voluntary wage restraint during the next two years and inflation was cut to 12 per cent within a year and to 8 per cent by 1978.

The social contract began to crumble in November 1977 when the Fire Brigade Union (FBU) called a strike in support of a 30 per cent wage claim. The army was called in to provide emergency fire services and the firemen, both for reasons of conscience and because they knew that if the strike caused deaths they would forfeit all public sympathy, did not interfere. Though damage to property and insurance losses were heavy, no deaths were attributed to the strike. Many firemen, however, were extremely unhappy about striking as well as with losses of pay amounting to some £500 each. The FBU delegate conference was recalled on 12 January and voted by 28 729 to 11 795 to accept the employers' 10 per cent offer. They did, however, extract one valuable benefit from the strike – a guarantee that their pay would be fixed at the top of the second quartile of manual workers pay. Surprisingly they were not required to accept a no-strike clause in exchange for this guarantee.

The full union revolt gathered momentum during 1978, when many shop stewards rejected the call of the Government and of their own leadership and the annual TUC conference voted against a third year of wage restraint. This rejection exploded early in 1979 in a damaging series of public service strikes – the 'Winter of Discontent' (see Chapter 3). Once again, this caused many to be laid off work (235 000) and considerable public suffering and, for the second time in six years, convinced the public that they had a Government which was unable to control union power. Once again, it was a public service strike – this time a 'coercive economic' rather than political strike – which brought down the Government in 1979.

MRS THATCHER 1980–3

Mrs Thatcher's Government, as described in Chapter 3, moved cautiously in its legislation to meet the insistent public demand for greater control of union power. She showed more finesse than any of her predecessors (Heath, Wilson and Callaghan) in handling public service strikes, and her task was made easier by high unemployment, which enabled her to sit back and let things take their course, applying only the sanction of refusing to give the public service managements more money. The British Railways Board (BRB) felt obliged to introduce flexible rostering to save manpower (and to keep within their budget) and the train drivers union (ASLEF) called two strikes, the first a series of 'guerrilla strikes' (17 one-day stoppages) in January and February 1982 and the second a full stoppage from 4–18 July. The Government did nothing and reiterated that the BRB would get no more money. The drivers, aware that they had each lost some £300 already, began to drift back to work, and ASLEF called off the strike with virtually nothing to show for it, but with badly damaged relations with the other manual railway union (the NUR) and an almost total forfeiture of sympathy amongst the travelling public.

During the same year, the two main Health Service Unions (COHSE and NUPE) called a prolonged series of intermittent one-day strikes over the period May to December 1982. In this case they had much more public sympathy, particularly from patients to whom the dedication and low pay of hospital staffs is very apparent. The strikers were careful to avoid alienating this sympathy by action which could cause death or serious suffering, though the chief sufferers were and still are those on the waiting list for non-urgent operations (e.g. hernias or replacement of hip joints). The waiting lists in some areas were doubled and the potential waiting time increased by more than 12 months. Banking on the underlying public sympathy the TUC called a national one-day strike on 22 September 1982 (though support was patchy). The Government however, did nothing other than reiterate that no more money would be made available and after 7 months the union had to abandon the strike, amid much bitterness and recrimination and virtually nothing to show for it.

The following year Mrs Thatcher had her landslide reelection and, six weeks later, picked the Health Service for the introduction of what will almost certainly be her approach to industrial relations in the public services as a whole. On 27 July 1983 she announced that there would be an independent pay review body for nurses and other professional medical workers but that any groups which resorted to industrial action would be excluded from the scope of the review body's recommendations. The Royal College of Nursing, with 226 000 nurses, which had always refused to take part in industrial action, welcomed the proposal (which had in fact been made as a result of an approach by them). COHSE (with 140 000 nurses) and NUPE (90 000), who had led the 1982 strike, condemned it.[16]

Next day the SDP leader, Dr David Owen, suggested in Parliament that if there were a ballot of all health workers (over one million) they would vote overwhelmingly in favour of a no-strike agreement in exchange for a fair method of assessing their pay.[17]

THE WAY AHEAD

Dr Owen's suggestion probably indicates the best of all ways ahead, initially for essential services and perhaps eventually for all public services. Legislation to ban public service strikes, or to hedge them with requirements for a cooling off period, secret ballots etc, has had a patchy record of success, as proved by the fate of the Industrial Relations Act of 1971 and by the experiences under the Taft Hartley Act in USA.[18] Far more effective would be to introduce a no-strike clause for essential services, strictly in exchange for the monitoring of pay by a pay review board as for the army and the police. It would, however, be far better introduced by consent than by legislation.

Such an arrrangement would have to be developed gradually for a number of reasons. It would be implacably opposed by the trade unions, especially at shop steward level, because it would remove what they see as their only means of exercising power and holding the support of their members – by successful collective bargaining to improve wages and conditions. It would also be extended with great caution by governments, because of the enormous proportion of the population who work in public services. Repeated settlements to match wages to prices for all of these could cause leap frogging and fuel inflation. Conversely, if workers in the private sector were to get bigger rises by industrial action, not only would they feel encouraged to do so, but their brothers in public services would get increasingly discontented. Earnings (rather than basic wages) tend to rise faster than prices in most years.

Nevertheless, introduced gradually, it should not only work, but could transform the entire system of British industrial relations because others would be attracted to the idea. People in general do not like striking and many would welcome a less disruptive way of getting better pay and conditions.

This indicates a new role for a very different kind of trade union representative. To avoid instigating inflation, the pay review bodies should not be bound by rigid rules connecting wage settlements directly with prices or with average earnings, but should instead be guided by the principle of equity (which is not the same as equality). This would mean that, at the end of each wage agreement, the union representatives would put their case before the pay review body for the next round, based on such factors as the value of the job, working conditions or unsocial hours – as they now do when a case is referred to arbitration or to a Court of Inquiry. They would act as advocates rather than negotiators and the skills required (which the best ones already have) would be to gather together facts and arguments with which to convince rather than coerce. Unions could, if they chose, use their funds to engage professional counsel to put their case. Even the most expensive QC would cost the members a great deal less than the £300 per head which is commonly forfeited by each striker in a two week strike.

Certain safeguards would be essential. Both the composition of the review bodies and their findings must be subject to appeal to a higher court and, if there is a case for it, right up to the House of Lords. Appeals, following the same principle as in civil or criminal cases, would be concerned not with the substance of the case but on the equivalent of the law, i.e. whether the review body had abided by its terms of reference and had been impartial.

Dr David Owen has suggested the secret ballot as the best method of introducing such a system. This should be applied initially to essential services (e.g. ambulance, hospital, medical, fire, electricity, nuclear power, gas, water, sewerage, postal and telephone services.) The secret ballot should ask each member whether he or she would wish to remain in the service if this required signing a no-strike clause in exchange for a pay review body with clearly defined terms of reference. It would be stated on the voting paper that, if a majority voted in favour, this would become a condition of employment in, say, two years time. Thereafter those who still declined to sign the no-strike clause would be discharged with normal severance payments, and no others would be recruited without signing it.

Policemen and servicemen have always accepted as a condition of employment that they cannot strike and there has seldom been any serious shortage of recruits. Many, in fact, give 'freedom from being called out on strike' as one of their prime reasons for joining their service.

One further safeguard would be desirable for all public services, with the possible exception of a few providing the 'essentials of life'. They should, every five years, be asked to vote in a secret ballot whether they wish the no-strike pay-review agreement to continue. (The two must go together. The guaranteed wage level for firemen in 1978 without a no-strike clause was a great mistake). The regular voluntary renewal by secret ballot and the rights of appeal would not only let justice be seen to be done; they would also give review bodies an incentive – indeed an imperative – to ensure that their awards really were fair.

Most workers in essential services intensely dislike striking, especially those who have regular contact with the public. Many can generally only be led to do it if their trade union representatives insulate them from the members of the public they serve at the time of the decision and thereafter during the dispute. Few would take such a decision personally, face to face with those depending on them. No ambulance driver would leave an injured child by the roadside and few would refuse a personal request to take a sick person to hospital. Few firemen would personally refuse to fight a fire if they were actually there. The vote to strike is taken in a vacuum. The postmen showed in 1971 how reluctant they were to carry through the refusal to deliver mail. People with this attitude to their work deserve a guarantee that they will not suffer for it. Those who have less direct contact with the public, such as electrical power workers and train drivers, are more ready to strike. Miners have the worst strike record of all, despite being paid by the public to mine a publicly owned commodity used by the public, possibly because they seldom meet people other than in their own community.

There is little doubt that many others who serve the public face to face – bus drivers and 'frontline' civil servants (e.g. in social security offices) even though not classed as 'essential to life', would also gladly forego the right to strike in exchange for a fair pay review system.

Nevertheless, an alternative means of getting fair pay alone would not be enough. Many strikes, particularly the small local ones, are concerned not with pay but with conditions – e.g. lack of ventilation or unfair treatment of a fellow worker. There must still be provision for resolving such conflicts in public services, either by permitting such strikes within clearly defined parameters (e.g. not in breach of agreement and not prejudicing the continuance of essential services) or by establishing a fast and efficient system of district tribunals by whom such minor disputes could be resolved. Though this would be expensive, it would cost far less in the end than the denial of a public service on which other economic activity depends.

If this path were followed, the idea might spread beyond the public services. Already a number of firms and unions have signed voluntary no-strike clauses for the duration of a wage agreement and this is much more common in USA than in UK. Just as most civilised people eschew violence as a means of solving problems so others may wish to eschew strikes. The Labour Government 1969 discussion paper was well named *In Place of Strife*. The trade unions would not lose their role; they would change it, becoming advocates to convince a third party, an impartial court, rather than negotiators to coerce. There is nowhere this could better start than in the essential services and Dr David Owen is undoubtedly right in his belief that the overwhelming majority of the Health Service would vote for it in a secret ballot. So would many others.

Part III
Communication and Participation

12 The John Lewis Partnership and Baxi[1]

A PARTNERSHIP OF 25 000

Of all the partnerships and cooperatives in Britain the largest, most successful and most original is the John Lewis Partnership. It has total assets of about £330 million (1983) of which £20 million is in fixed interest preference shares and debentures, the remaining £310 million belonging to the 25 000 employees collectively all of whom, from Chairman to cleaner, are therefore Partners. The share capital is not issued to individual Partners but is held in trust for their benefit while they are actually working for the Partnership; in other words, they contribute no capital when they join and cannot take away any when they leave.

The Chairman holds his position subject to the continuing approval of two-thirds of the members of the Central Council acting through three Trustees elected by it. At least 80 per cent of the Council are elected by universal suffrage of the Partners. The Council thus holds the ultimate power over the continuance of the Chairman in office but there are strong constitutional restraints (see below) to guard against this power being used hastily. The Chairman, who is also Chairman of the Trust Company, enjoys considerable independence in commercial decisions but only so long as he retains the support of two thirds of the Central Council and the elected Trustees, in default of which they have the power to dismiss him.

After the deduction of prior calls on profits (tax, non-contributory pensions, numerous amenities, etc.) the balance is divided each year between reinvestment for development and a cash bonus for distribution to the Partners. Over the past 10 years an average of about 60 per cent of the profit has been ploughed back and 40 per cent distributed, though in 1982–83 the cash bonuses (17.1 million) exceeded the reinvestment (£16.3 million). Estimated results for the year were as follows:[2]

TABLE 12.1 *John Lewis Partnership Results 1982–83*

		£ million
Total	Sales	921.8
	Trading profit	50.6
Prior Calls:	Corporation tax	5.1
	Dividends on preference shares	0.4
	Interest on loans	6.0
	Contribution to pension fund	5.7
Balance for Disposal		33.4
Disposal:	Partnership bonus	17.1
	Retained for expansion, development and inflation	16.3

This £17.1 million bonus was paid in the form of individual bonuses amounting to 16 per cent of each Partner's pay. This has been the average annual bonus over the past 10 years, the highest being 24 per cent in 1979 and the lowest 13 per cent in 1975 and 1976. At the bottom of the recession (1981) the bonus was 14 per cent.

The John Lewis Partnership has unarguably been a commercial success. Since its foundation in 1929 its capital value has grown from £1 million to £320 million. In real terms this growth is 23 times greater than inflation. Few firms can better this. The numbers employed have also grown. Their earnings match those of others in the retail trade and their bonus takes theirs well above that. Industrial conflict is almost unknown, redundancies have been and remain small. Though there is some criticism of the effectiveness of the democratic control (see below) the partners undoubtedly have more means of influencing the operation of the business, and their own remuneration and conditions of work than most employees in firms of their size. The result is a stable and generally happy work force which has enabled the firm to maintain its profitability and growth through two major recessions (the 1930s and the 1980s).

EVOLUTION

The Founder of the John Lewis Partnership was John Spedan Lewis. His father, John Lewis had opened a draper's shop in Oxford Street in 1864. He later acquired the derelict business of Peter Jones in Sloane

Square which, in 1914, he made over to his son Spedan (already a partner with him in John Lewis) with a free hand to 'do whatever he liked with it so long as he did not leave the Oxford Street business before 5 o'clock in the afternoon'. Peter Jones was at that time making a loss but Spedan galvanized the work force by telling them that, when the firm made a profit, they would share it. In 1920 he was able to pay them the first 'Partnership Benefit' which amounted to seven weeks pay in promissory notes over and above their earnings for the year.

Spedan Lewis inherited the whole business on his father's death in 1928 and in April 1929 he made his first settlement in trust for the benefit of the workers, present and future, in John Lewis and Peter Jones. The process was that he sold to this Partnership the whole of his right in the two businesses at a very conservative estimate of their market value (about £1 million) but left the money due to him as vendor in the hands of the Partnership as an interest-free loan with the provision for its repayment over a long period. In 1950 he made a second settlement, which was irrevocable and at the same time established the Constitution, the Central Council and the system of control described below.

In the 1930s, the Partnership had acquired a number of additional department stores and also extended its activities into food retailing. By the early 1940s it owned 18 department stores, 33 food shops and 5 other specialist shops, employing over 11 000 men and women as Partners.

By 1983 it owned 20 department stores and 74 Waitrose supermarkets employing 25 000 Partners. The largest store is John Lewis in Oxford Street with over 2000 employees. Six others (including Peter Jones) employ over 1000 each. The average size of the staff in each of the 74 Waitrose supermarkets is about 100. Two thirds of the 25 000 Partners are female.

EXECUTIVE STRUCTURE

The essential feature of the John Lewis Partnership is that it is run on straight commercial principles by the Chairman and a board of directors but that he is answerable to and, in the last resort, subject to dismissal by the democratically elected councillors. This seldom creates problems since the Partners share the same aim – to make a profit for development and distribution and to enrich their working lives.

The Chairman is appointed by his predecessor, though the new incumbent could be dismissed if Central Council so voted by the required majority. When Spedan Lewis retired in 1955 he appointed a successor who had worked for 28 years in the company, Mr (later Sir) Bernard Miller who in 1972 appointed Spedan's nephew Peter Lewis, the present Chairman.

Unless the Council exercises its 'massive retaliation' the Chairman enjoys overriding power. He chairs the Trust and also the Central Board and appoints the executive directors. These directors run sections of the business – e.g. the Director of Trading, Department Stores (20 stores), the Director of Trading, Food (the 74 Waitrose supermarkets), and the Directors of Personnel and of Finance. This is a normal executive structure of branch heads, directors and managers.

The ideals of the Founder are recognized and respected by the Partners, but senior executives and most of the work force know well enough that maintenance of these ideals depends on the business making a profit and, if possible, expanding to ensure that there are enough jobs to absorb redundancies arising from closure of outmoded operations or from new technology.

Standards of management are high. Competition for places as graduate management trainees is keen and those selected are all thrown in at the deep end as sales assistants from September to the end of January i.e. through the Christmas rush and the January sales. The personnel department are wary of any who are too 'starry-eyed' and aim to recruit managers who, they judge, will do the work within the Partnership ethics. Within a year they may be section managers and the good ones can become departmental managers in 2 or 3 years, some moving across to buying or other departments.

Thereafter managers are judged by results which are published for all to see. In the Partnership's *Gazette* the turnover of the department store and food divisions are published each week, together with the percentage changes over last year of every buying section, every department store and every Waitrose supermarket *with the name of the manager*. Similarly, in each branch newspaper, in its own *Chronicle*, the performance of each department is published with the name of the departmental manager. This leaves no doubt about the chain of responsibility and encourages managers at all levels to insist on decentralisation, which is at the heart of direct and personnel management.

Discipline is thus built on sure foundations of commercial performance for the Partnership at every level. The checks and balances of the participatory structures keep problems to a minimum and can usually

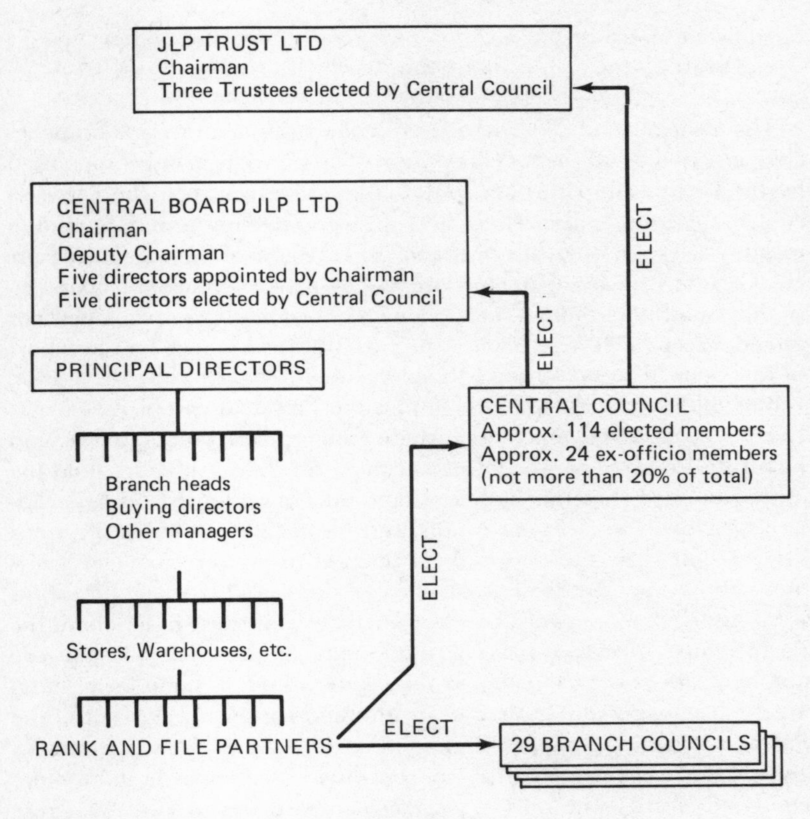

FIGURE 12.1 *John Lewis Partnership (JLP)*

resolve them. There is a standard of efficiency which reflects the high morale and confidence – and this factor must certainly have weighed in the mind of Spedan Lewis as a successful businessman alongside his equally genuine philanthropic ideals.

THE PARTICIPATORY STRUCTURE

The first pillar of control of the Partnership is the Central Council and, as is usual in such structures, having been given the power, it uses it with restraint and responsibility. To quote Spedan Lewis again:

> Human institutions so devised that they cannot work badly may be so encumbered with safeguards that they cannot work to any

sufficient purpose at all. . . . Organizations, no matter how elaborate, come at last to a point at which there has to be trust.

The Central Council varies in size around 140 members, as branches close or expand. Of these, a maximum of 20 per cent may be appointed by the Chairman. These are, in practice, the executives who can best contribute to the discussions and answer cross-examination, and a rotating selection of Heads of Branches, Directors of Buying, Registrars etc. Certain Principal Directors are amongst the ex-officio members. So is the Chairman though he, following Spedan's example, does not attend, except once a year when the Constitution requires him to report to the Council in person and to answer questions. In 1983 there were 138 members, 114 elected and 23 plus the President ex-officio.

The 114 elected Councillors include a high proportion of middle and junior managers. The previous Chairman, Sir Bernard Miller, told the author in 1970 that this bothered him and, as he toured his branches he urged them to elect more rank and file members but the Partners argued that their own store, departmental or section manager knew more about their needs than anyone else and 'can't we elect him if we want him?'. A more cynical view, sometimes expressed in letters to the Partnership's weekly *Gazette*, is that many of the rank and file are apathetic about the Council and that those willing to stand for election are often unopposed.[3] In fact, branches representing 25 per cent of the Partners returned unopposed candidates in 1982 and 44 per cent in 1983. Of the 14 000 Partners in constituencies in which there were contested elections in 1983, 73 per cent turned out to vote. Waitrose constituencies averaged 86 per cent turnout, the highest being 94 per cent. Some of the larger department stores showed a lower turnout (e.g. John Lewis in Oxford Street 45 per cent) or returned candidates unopposed (e.g. Peter Jones).

The preponderance of managers amongst the candidates elected or returned unopposed to the Central Council continues: 57 per cent in 1982 and 62 per cent in 1983. Most of these were selling department managers, section managers and Waitrose branch managers and, of the 43 rank and file councillors in 1983 only 15 were department stores selling assistants.[4] This does, however, suggest that there is far less 'them and us' feeling than in most large companies and if the rank and file Partners wished to challenge or confront the management they would use their power to elect more rank and file members to do so. Once again, given the power, most people prefer to use it constructively.

The underlying atmosphere in Council debates is certainly constructive though, as will be shown later, it can be both critical and heated on occasions.

Each branch also has a Branch Council which has a similar role and, some consider, one which more directly affects the day to day work of the rank and file, so there are few signs of apathy here. There are 29 Branch Councils including one for each of the 20 department stores, a big one covering all the 74 Waitrose supermarkets and others covering various specialist and service departments and Central Offices. A branch Councillor can also be elected to the Central Council and in 1983, 49 per cent of the elected Central Councillors were also Branch Councillors.

The Waitrose branch elects 22 of the 114 Central Councillors, the biggest department store (Oxford Street) elects 7 and the smallest elect 2 each. They hold office for one year (March to March) but remain eligible for reelection. In practice, some choose to take a break and stand again later, so that, of the elected Councillors in 1983 a little over half had served in the previous year and 29 per cent were serving for the first time.[5]

The second pillar of control is the Trust Company – or to give it its full title – John Lewis Partnership Trust Limited. which is legally the chief trustee of the two settlements originally made by Spedan Lewis in 1929 and 1950. These settlements are now irrevocable and can be modified only by the consent of a court of law as far as the law allows. In essence, the 1929 settlement provides for the division of all the profits among the workers after certain prior charges have been met. The establishment of John Lewis Partnership Trust Limited in 1950 and the 1950 settlement provide the authority and constitution under which the Partnership is managed, and the Trust Company is responsible in law that both settlements are carried out and that the management does act within the Constitution. The Trust consists of the Chairman, the Deputy Chairman and three Trustees elected annually by the Central Council. These three Trustees can, as described above, dismiss the Chairman if two thirds of the Central Council so demand.

It is here that the most significant safeguard in the Spedan Lewis Constitution operates: until and unless that two-thirds majority of the Central Council votes no confidence in the Chairman, the three Trustees have no voting power in the Trust Company other than certain limited rights on a resolution to wind up the Trust Company or alter voting rights. The machinery for this is that the Trustees hold 60 per cent of the nominal £100 of shares in the Trust Company but these are non-voting shares except as indicated above. In the event of the Central Council passing a two thirds majority vote of no confidence, the

Constitution provides that the Trustees' shares would acquire voting rights which they must then exercise in accordance with the Central Council's wishes. In other words, it is the Central Council which holds the real power to dismiss the Chairman but, unless it exercises that ultimate power, the 'massive retaliation', the Chairman himself holds the power of control of the Trust Company and, through this the power to direct the management of the Company, appoint executives, open new stores or close old ones – in fact, to run the Partnership.

This is probably where the genius of Spedan Lewis lay, because in it lies the germ of the commercial success of the Partnership from which all else has flowed. The Chairman's position is, in many ways, more powerful than that of the Chairman of a Joint Stock Company, who can be overturned by the Board of Directors elected by the shareholders. Some critics argue that the John Lewis' Chairman's power makes the Partnership's democratic control illusory, but the fact remains that he does only hold this power subject to his retaining the support of the Councillors elected by the shop floor so that is where the power ultimately lies; and Spedan Lewis gave this power, irrevocably and of his own volition, when he was the sole share holder and therefore immune from dismissal by anyone but himself.

Spedan Lewis was above all, a strong and confident man who believed in strong management, and he clearly knew that the Partnership would need to be able to survive problems with the future Chairmen, less strong and with less prestige and, perhaps in times of recession, facing embittered workers fighting for their jobs. Between 1929 and 1950 he, and they, had experienced both the Depression and a world war. His 1950 Constitution ensures that the Chairman will retain the power to manage with commercial success without looking over his shoulder and the Partnership's record in the recession of the early 1980's bears out its success. As will be shown in Chapter 13, the lack of strong management has been the major cause of failure in other attempts to run cooperatives.

So this second pillar of the Partnership is in practice the Chairman himself rather than the elected Trustees, who have never yet acquired their reserve power which the Central Council, the first pillar, can give them, and probably never will.

The third pillar is the Central Board. This board, in many ways, has more in common with a German Supervisory Board (see Chapter 4) than with a German Executive Board, in that day to day management of the John Lewis Partnership is conducted by the individual Principal Directors, who also meet regularly to coordinate and discuss matters of common interest. Five of these Principal Directors are members of the

Central Board, and there are also five directors elected by the Central Council. The management does, however, in this case appear to have control since the Chairman, his Deputy Chairman and five nominated Principal Directors have a majority of 7 to 5 but in practice the Board acts as a unity and differences of opinion are not rooted in whether a particular Director is nominated or elected. In essence the Central Board controls the finance which guides and constrains the policies which the Principal Directors carry out. The five elected directors do, however, have certain special powers; if, for example, three of the five so require the Board is obliged under the constitution to consult the Central Council on any decision involving redundancy of more than 12 Partners or contraction or expansion involving more than 5 per cent of fixed assets (other than by accumulation of profits). In practice, however, the powers of all directors are equal, no matter by which route they became directors. The details of the workings and relationships between the three pillars – the Chairman, the Central Board and the Central Council - will be illustrated by an example of a conflict later in this chapter.

COMMUNICATION

Another key element in participation is the system of Committees for Commmunication run by the Partners' Counsellor. He is a senior Principal Director (the previous one is now Deputy Chairman) and is a kind of ombudsman. He can be approached direct by any Partner over any problems, personal or general, and he, of course, refers directly to the Chairman. He runs Committees for Communication at each branch, whose members must themselves be rank and file (i.e. without any management responsibilities) elected by rank and file. They meet four to six times per year under a representative of the Partners' Counsellor. Like Works Councils in Germany (Chapter 4) and Communication Groups in Marks and Spencer (Chapter 14) they can and normally do invite the head of the branch to attend part or all of their meetings, so that questions and criticisms can be raised. That is done by the Partners Counsellor's representative to preserve the anonymity of the committee members. All questions and discussions are minuted and published in the branch *Chronicle* but free discussion is encouraged by not recording any names in the minutes.

Amongst the most impressive manifestations of free discussion is the Partnership weekly *Gazette*; which was first published in 1918. While serving the normal house magazine functions (news, trading results, recreation, etc.) it has full reports, on *Hansard* lines, of Central Council meetings. It also has a lively and often outspoken correspondence column, whose strength lies in the fact that correspondents are positively *encouraged* to be anonymous. With each is printed a reply by the management (usually a Principal Director or the Chairman). The result is an expression of free speech of which the author has seen no parallel – some examples are given under 'The Constitution at Work' later in this chapter. A remarkable feature of the *Gazette* is that it is available for sale not only to Partners (at a nominal price) but also to the public at large. To read it regularly will convince anyone but the most prejudiced of sceptics that the democratic process in the Partnership is genuine and that the management, so far from stifling criticism, gives it full publicity and responds to it. The General Editor has a high degree of independence and is a Principal Director, a fact which further underlines the importance attached to this channel of communication in the health of the Partnership.

THE CONSTITUTION AT WORK

It is very rare for the Chairman to reject a recommendation by the Central Council[6] but he may do so, after reference to the Central Board. In practice this means that he and the Principal Directors consider that its acceptance would have serious commercial consequences. The Chairman then announces and explains his decision, either by a statement at a meeting or (to give more time for digestion) by giving his reasons in the *Gazette*.

On 20 September 1982 the Central Council recommended a change in the Spring holiday regulations which, after discussion with the Central Board, the Chairman turned down. In essence, current rules required that at least one week of the annual holiday entitlement must be taken between November and the end of April. The purpose of this was to avoid the work force being too heavily reduced in certain popular holiday months. The Central Council resolved that May should be added to the months in which the Spring Holiday could be taken. The members from the biggest stores (Oxford Street, Brent Cross and Peter Jones) mostly voted for the proposal and the smaller ones (especially Waitrose) against it. The final vote was close – 58 to 46 with 9 abstentions, though it must be assumed (the ballot being secret) that most of the 24 ex-officio members voted against the proposal so the vote of elected members may have been more than 2 to 1 in favour.

The Chairman and his Principal Directors – particularly the two Directors of Trading – felt strongly that the original reason for this restriction remained valid; the proposed change would seriously distort the balance by further increasing the holiday absences in the summer, since few could be taken in November and December and the unpopular months of January, February and March would be even more undersubscribed if Partners could count a week in May as their Spring Holiday rather than as part of their summer holiday. After discussions in the Central Board the Chairman published a 2500 word statement in the *Gazette* four weeks after the meeting (23 October) explaining why he could not accept the proposal as it stood but saying that he would apply it for a trial period of two years in the two largest branches which had pressed for it most – Oxford Street and Peter Jones – and in Brent Cross which has different and complicated trading and staffing hours. In his statement he was openly critical of the management of the Oxford Street branch for failing to exercise proper control over distribution of holidays.

The result was a furore, judged by letters to the *Gazette*, not so much about the substance of the case as over the principle that the Chairman had exercised his right of veto over a matter which did not seem to many of the Councillors or the rank and file to have serious commercial consequences. With hindsight, it seems probable that the Chairman's decision needed only to be put more diplomatically – e.g. 'I believe that it will be best to introduce the Council's proposal by degrees rather than at once. I propose to try it out for the next two years in the three largest stores. This will give us time to evaluate how well it works and amend it if necessary before considering its extension over the whole Partnership.' But diplomatic writing consumes time which the Chairman may not always have,[7] and the way he put it clearly did strike some raw nerves.

For the next six months the controversy raged in the Letters Column of the *Gazette*, most correspondents observing the traditional anonymity and the Chairman, his Deputy or a Principal Director replying in print to each batch of letters. In particular, the provisions of the Constitution and the role of the Chairman were challenged e.g.

> I find your comments pointedly aimed at department managers in Oxford Street quite unjustified, scurrilous and downright demoralizing. . . . it seems to me that we are now going to use a

sledgehammer to crack a very small nut . . . . The members of senior central management whose advice you say you have taken have absolutely no idea of the day-to-day problems that exist on the selling floor in this day and age. . . . The *us* and *them* brigade have had a field day with your comments. You have done more harm than you could possibly comprehend by your action. . . . (30 October 1982)

It would appear that our experiment in industrial democracy is now over. We can safely announce that it has *not* worked. (30 October 1982)

So much for democracy! (6 November 1982)

Perhaps the Chairman is unaware of the growing indifference and cynical attitudes that prevails at constituency meetings, the constant remarks such as, 'Oh, what's the use' or 'The Council is nothing but a rubber stamp to be used on management decisions.' (6 November 1982)

Whatever happened to the Constitution? The Chairman says he will not accept the Council's decision for the great majority of Partners unless 'it would plainly be to the Partnership's commercial advantage'. The rules say he should let the Council decide unless to do so would 'cause the Partnership injuries so great that they should not be risked '. (6 November 1982)

. . . . he has made an utter mockery of both Branch and Central Councils. (6 November 1982)

I would like to know the views of my fellow partners on the proposition that our next Chairman should be elected and not appointed. (13 November 1982)

I shall never again bother to vote in branch and central elections. What a waste of time. Why does anyone bother? . . . . Why not disband the Councils and stop this pretence?. (20 November 1982)

If the Central Council were to vote (with a two-thirds majority) to oust the Chairman, would he have the power to veto this decision too?' (The General Editor replied 'No' and explained the constitutional difference). (27 November 1982)

The above quotations alone may give a misleading impression without the equally forthright replies by the Chairman and directors, and other letters from those who supported them. The purpose of the quotations, however, is to demonstrate the reality of the *Gazette's* free speech. The fact that it is encouraged, anonymously, and printed gives strong proof of the confidence and strength of the Partnership.

At a later meeting of the Central Council on 7 February 1983 the seconder of the original resolution introduced a further proposal:

> That this Council requests the Chairman that, should he decide in future to refuse or only partially accept this Council's recommendations such a decision should be accompained by the fullest and quantified statement of the reasons why the Partnership would be put critically at risk by this full acceptance.

The proposer argued that, while she accepted that the constitution did give the Chairman the right to reject such a recommendation, he was not bound to give any reason if, after consultation with the Central Board, it was decided that it would be 'inexpedient' for the Partnership to accept. She wished to bind the present Chairman and his successors to give a factual and detailed report.

Against the proposal it was argued that the Chairman *had* given full reasons, in his long statement in the *Gazette* on 23 October and in his answers to readers' letters in subsequent issues; that to tie his hands commercially would be a retrograde step and he could not in any case bind his successor; and that if a Chairman continually made judgements which were unpopular and unreasonable, genuine public opinion would bring itself to bear and would make the Council take the necessary action to get rid of him. No one, and certainly not the proposer, was suggesting this.

In a debate conducted with considerable restraint the general feeling was best expressed by a comment that the Chairman and Principal Directors 'had not handled the issue with their accustomed dexterity . . . and for the subsequent hassle had only themselves to blame' but that it had been blown up out of proportion to its importance. In a secret vote the Council rejected the proposal by 73 to 34 with 8 abstentions.

Meanwhile, at the annual Central Council Dinner on 22 November the Chairman had floated the idea that 'instant confrontations' in the Council before management had had time to consider the implications of the proposal should be avoided. This might mean a change in Council procedures. He expanded on this in an article in the *Gazette* on 29

January 1983 in which he suggested that, if a proposal to change a significant commercial practice were resisted by the Principal Director concerned a first and second reading procedure should be adopted so that the proposal, possibly modified to meet problems brought to light in informal discussion in the meantime, could be debated at the next Council meeting when management would be better able to present considered comments on the implications.

This proposal was made the subject of a 'debate on the adjournment' at the Council meeting on 28 March, a procedure which allows free discussion of the broad issue without a specific motion and without a vote. This enabled various methods of procedure to be discussed, e.g.: that the decision for a second reading should be taken by a majority vote by the Council; or if a Principal Director so requested during the debate; or if the General Purposes Committee of the Council so decides, or the Central Board so requests before the debate; or that a proposal passed by the Council should be returned for a second reading if rejected by the Chairman.

These ideas were examined by the General Purposes Committee which put forward a proposal to amend the Standing Orders of the Central Council. The new Standing Orders provide that during the course of a debate on a motion that is not already subject to the first and second reading procedure, a member can propose that it should be returned to the Council for a second reading; if this is seconded and carried by a majority vote then the business under discussion will continue until a vote to take note of the substantive proposal is taken. It also provides that if a first reading is accepted by more than one third of those present voting in favour, then the proposal (or the amended proposal) shall return to the Council for a second reading at some subsequent meeting.

This amendment to Standing Orders was passed by the Central Council on 23 May 1983 by a majority of 127 in favour, 1 against and 2 abstaining in an open vote.[8]

CAN THE JOHN LEWIS EXPERIENCE BE REPEATED?

The Deputy Chairman of John Lewis Partnership commented in a letter to the author in April 1983:

> One of the central difficulties in setting up this sort of organisation is only rarely mentioned – it is the fact that while rich men may

give large sums away with varying degrees of publicity, they are almost never willing to give away power and that in the end is what Spedan did.

It has probably been easier to make this work – and maintain the constitutional safeguards which have ensured strong management and commercial success – in a retail business than in the manufacturing industry. The workforce is predominantly female, which leads to a relatively fast turnover (due to marriage, childbirth and changing family commitments). Two thirds of the John Lewis Partners are female, and the total labour turnover is about 17 per cent per year. Though they are free to join trade unions, very few do.

The same factors make it easier to cope with the recession and redundancy, most of which can be absorbed by the high natural wastage of female employees.

The 1982 edition of the Partnership's booklet *About the John Lewis Partnership* ends with the words

> Indeed, it seems entirely possible that, in a society that is beginning to stir purposefully in response to the idea that gainful occupation should offer something more than gain, the example of the Partnership may justify the ideas of John Spedan Lewis over an area much wider and more varied than the retail field – and this would be the culmination of his dreams.

The following year there was a development which could prove to be a major step in that direction.

BAXI

The view that the John Lewis pattern of partnership could not be applied to manufacturing industry was challenged head on by Mr Philip Baxendale, Chairman of Baxi Heating, in March 1983.

Baxi was originally founded by Mr Baxendale's great-grandfather in 1866 and has remained a family firm ever since, manufacturing a series of successful stoves, the most famous of which have been the Baxi back boilers which fire a central heating systems at the back of a domestic solid fuel or gas fire. By 1983 the firm was employing 900 people with a £37 million turnover and a £5.8 million profit of which all but £90 000 was ploughed back into the business.

Such a small but successful business was clearly attractive for takeover bids and Mr Baxendale, by then 56 decided that, if he sold the business on the open market, either one of the giants would take it over by buying the majority holding or the new shareholders would no longer be able to resist the lures of takeover bidders.

He therefore followed a similar course to the one pioneered by Spedan Lewis in 1929 and 1955. He and the other shareholders sold the entire business – estimated to have a stock market value of over £21 million – to an employee trust for £5.2 million. The Trust has three trustees, one of whom is Mr Baxendale. As with John Lewis the company is run by an Executive Board responsible to an elected Partnership Council.

One important difference is that Baxi will, over a period of years, pass 49 per cent of the shares to the employees under participation schemes which they intend will be approved by the Inland Revenue. Thereafter, employees will be free to sell their individual shares to the Trust and a controlling 51 per cent at least will always be held by the Trust on behalf of the employees actually working for the company.

Partners of the Baxi Partnership will be free to join trade unions and these unions will be recognized for negotiation with the Executive Board.

It is still too early to say how well the Baxi Partnership will work but initial prospects were excellent, with profits rising despite the recession, and expectations of further rises as the recession passes. The 900 strong work force, which had been participating in a works council and a cash profit-sharing scheme since the 1960s, greeted the new Partnership proposals with enthusiasm. If expectations are realized, this could be the breakthrough for extending the partnership principle into a wider range of manufacturing and service industries.

OTHER POSSIBLE VARIATIONS

Whether Baxi will succeed and grow like John Lewis remains to be seen, but the fact will remain that apart from Spedan Lewis and Philip Baxendale few have been willing to give away both money and power and it would be unrealistic to hope that it will ever be more than exceptional. The commercial performance of industries nationalized by compulsory government buy-outs has been so poor – and their industrial relations generally so much below par – that this is never likely to achieve either the degree of participation or the success of the John

Lewis and, it is hoped, of Baxi. The experience of self-management in Yugoslavia might be difficult to transplant to a democratic society and does not seem to have prompted imitation in other one-party states.

Cooperatives are also discussed in the Chapter 13. These are usually small, are often born in desperation out of commercial failure and suffer from weak or inexpert management, though there are exceptions to all of these. One of the most famous cooperatives is Mondragon in Spain and this has the unusual feature of being 80 per cent financed by its Cooperative Bank.

This gives rise to another possibility in the field of Partnership. The success of John Lewis has been such that if an enlightened Merchant Bank had provided the £1 million capital which Spedan Lewis handed over, and had legally vested the power in the Chairman and the Partners exactly in accordance with his Constitution, they would have seen their £1 million grow to an extent rare even for merchant bankers. The doubt, however, must be whether the Partners would have reacted in the same way if the Trust capital had been owned by a bank. Would it have been any different in practice from the many businesses – some highly successful – which are already financed by merchant banks?

One variant might offer a profitable experiment for a Merchant Bank: to launch a Partnership with a constitution on the John Lewis/Baxi model with a contracted option for the Partnership to buy itself out after a specified number of years at a price which would give the Merchant Bank a better than average return on its investment taking account of inflation. This would give both the Partnership and the Bank an incentive to generate the necessary growth and profit and the John Lewis experience suggests that both should have an excellent chance of doing so.

Another variant would be for the Bank, as the Partnership expands, to retain a stake in the capital in the form of preference shares, receiving dividends at a rate rising faster than inflation, but not sharing in the capital growth. If the company prospered and ploughed back most of its profits, the time would come when the Bank's preference shares, though increasing year by year in yield, would amount to a dwindling minority of the Partnerships capital.

To function successfully, any such scheme would need a number of essential ingredients: the Partners must be confident that the majority ownership would within a reasonable time pass into the hands of their Trust and that democratic control on the John Lewis lines would operate from the start; and the Bank should have reason for confidence that the return on its investment should be at least as high as would be likely from any other good alternative investment for which it might have used the money. Given the enormous commercial advantages which follow from the Partnership principle, such confidence should be possible on all sides.

13 Cooperatives

COOPERATIVES AND COOPERATIVES

The cooperative idea is one of the relatively few connected with the British economy which is supported by all the main political parties. The Act setting up the Cooperative Development Agency (CDA) was passed by Parliament with all-party support in 1978, and had the task of providing finance and other support and advice to groups of workers who wished to form cooperatives. Mrs Thatcher's Government, elected the following year, continued and expanded the operation. In 1978 there were 150 cooperatives in Britain; by 1983 there were 900.[1] There is every sign that the movement will continue to grow, particularly in service industries, facilitated by the burgeoning of communications technology. With the availability of support from some of the many organizations described below (e.g. the CDA, ICOM, ICOF, JOL), the prospect of continuity arising from all-party agreement and the attraction of creating a new and self-controlled destiny after the trauma of redundancy is resulting in many people – particularly the more adventurous ones – being ready to pool their redundancy money in a small cooperative venture.

Such small scale ventures, whether in production or service industries, are genuine 'workers' cooperatives' – i.e. they are primarily owned and controlled, even if sometimes primed by external capital, by workers who have staked their own capital on their success.

There are, however, many other kinds of cooperatives. These include 'consumer cooperatives', controlled by the consumers rather than the workers. In practice this control sometimes differs little from that by the shareholders in a joint stock company (e.g. see 'The Co-op' below). Some larger scale workers cooperatives have succeeded where a central source of capital plays or has played a prominent role (e.g. see Scott Bader and Mondragon below). Others have failed (e.g. KME, Meriden, the *Scottish Daily News*, of which KME has been selected for a case study).

'THE CO-OP'

To many people in Britain, the word 'Cooperative' will conjure up a retail chain store on the high street, one of 200 independent retail societies loosely coordinated by the Cooperative Retail Services Ltd (CRS), supplied and supported by The Cooperative Wholesale Society (CWS), all of them (with other activities such as banking, insurance, travel and education) embraced in the Cooperative Union. These collectively are known informally as 'The Co-op'.

The CRS is the largest retail cooperative society in the world, with 2000 outlets and 28 000 staff. It is not, however, a 'workers' cooperative' – i.e. it is not owned or controlled by these 28 000 workers but by its 2 million members who are in fact its regular customers. They have, in effect, a very similar role to that of shareholders in a public limited company.

The origins of the CRS and CWS lie with the 'Rochdale Pioneers' who formed a genuine workers cooperative in 1842. The pioneers were much influenced by the philosophies of Robert Owen whose experiments in his Lanarkshire mill in the early 1800s and his subsequent fostering of the first trade unions led him to be known, deservedly, as the father both of British Socialism and British Trade Unionism. Because of these origins 'The Co-op' has always had close links with the Labour Party; though with the right rather than the left.[2]

While other influences such as the Christian Socialist Movement led along the line of workers cooperatives, with the formation of the Cooperative Productive Federation (CPF) in 1882, the retail cooperatives increasingly drew their capital in from their customers, attracted by the benefits of discounted prices in exchange for a small investment at low risk. In 1862 the CWS was formed and in 1870 the Cooperative Union, as a central organization for both the CWS and the growing number of independent retail societies.

In 1982 the Cooperative Union Congress resolved that the CWS should establish its own retail outlets, initially under the name of the CWS Retail Society, changed in 1957 to its present title of CRS. Over the years the Cooperative Union extended its activities into other areas such as the Cooperative Bank and the Cooperative Insurance Society. In 1980 it also incorporated the CPF.

Anyone over 16 can join his local Retail Cooperative Society for a minimum investment of £1, though he can buy up to 10 000 shares if he wishes. His £1 share gives him a vote in the election of the Board of Directors at the Annual General Meeting. The Board appoints Officials and Managers who in turn recruit the staff. The capitalization is partly from members' shares and partly from reinvested profits and bank loans. This is exactly as in a normal public limited company except that the Co-op member has one vote whether his shareholding is £1 or £10 000, whereas the PLC shareholder has a vote for each share.

In addition to a normal rate of interest on his shares (substantial for 10,000 shares but nominal for £1) the member receives a dividend on every purchase he makes. In early days this was recorded against his membership number and he collected his dividend at the end of the accounting period. Today, the dividends are paid in Co-op stamps at the time of the purchase.

As well as voting for the Board of Directors of his Society, the member also has a vote for a representative on the Regional Committee and the 20 Regional Committees in turn elect 6 of the Directors on the CRS Board, 6 more being appointed by the Board of CWS, the Cooperative Union.

So the most significant feature of the Co-op is that, having been formed as a socialist alternative in rejection of capitalism, it has reached almost exactly the same economic structure as that reached by the public limited company which evolved through the capitalist route: the advance of the enterprising individual proprietor, then a proprietor with his sons or with a business partner, then a group of partners who eventually surrendered their power to the votes of shareholders in a joint stock company, who can elect and dismiss their Board of Directors who can in turn elect and dismiss their Chairman – but who also appoint the managers and run the company. These two routes were each dictated by the realities of raising capital and of the overriding need for efficient commercial management answerable to, but not too much hampered by, a democratic body. Some people are sad or exasperated at this departure from the original socialist principles but it is probably inherent in the principle of the consumer cooperative as distinct from the workers' cooperative.

FOSTERING AND FINANCING SMALL COOPERATIVES

The idea of the small cooperative must be as old as civilization, when the first group of people discovered that they could gain more by pooling their resources and their labour than by operating alone. It is a great survivor and, for example, the small family business, classed officially as a cooperative, continued to provide over 70 per cent of China's retail trade through the most radical period of the 1960s and 1970s. The most difficult problems arise when the cooperators need to raise extra capital from outside, to start, to survive, to trade and to expand. Whether they borrow it from a bank, a benefactor or an institution, they risk losing at least some element of control. For this reason, many cooperators prefer to raise capital from institutions idealistically committed to the principle of workers' control, rather than from a bank.

One of the most successful of these institutions has been the Industrial Common Ownership Movement (ICOM) and its subsidiary Industrial Common Ownership Finance (ICOF). ICOM was founded in 1958 by a Quaker and Christian Socialist, Ernest Bader, who had already handed over his own successful chemical business to its work force in 1951 (see Scott Bader below). ICOM's aim is 'the achievement of democratic control and ownership by people at work of the enterprises in which they work'.

ICOM's principle is that the capital of the enterprise is collectively owned by the workers but that no individual worker should own more than a £1 share, though inevitably this cannot be applied to the varying amounts of individual capital put in by the founder members. Many ICOM cooperatives were converted from private firms for social rather than economic objectives and in these cases most of the capital needed was already there. An ICOM cooperative does, however, sometimes have to raise further capital by seeking loans from its members but when it does so it tries to draw on equal amounts from each one, to avoid giving any individual a disproportionate degree of influence. This is usually practical in the small concerns which make up the majority of ICOM's cooperatives, often in service rather than in manufacturing industries and comprising middle class people attracted more by a way of life than economic considerations.

Though always motivated more by idealism than profit, ICOM has needed gradually to extend its activities to help to infuse the capital its cooperatives need to survive. In 1973 it formed the ICOF for this purpose and in 1978 it tried to form close links with the ailing CPF before the latter was absorbed into the Cooperative Union in 1980. ICOM has, however, been criticised by the Banks, the Cooperative Union and the CDA for its stance on the avoidance of control by shareholders internal or external, and ICOM has in turn accused the Cooperative Bank of 'conservatism' in its reluctance to provide loan capital to cooperatives at high risk against few assets.[3]

The Government-sponsored CDA was described at the start of this chapter. Predictably its philosophy is to encourage cooperatives only if they are judged to have a chance of achieving commercial viability and is reluctant to get involved in 'rescue cases' such as KME, Meriden or the *Scottish Daily News*.[4] It has no funds of its own (other than for its own administration and research) and so acts rather as a catalyst for raising funds from other bodies including government sources and local councils, and it also offers advice on organization and legal matters. Because of its pragmatism it is regarded with reservation by some of the more idealistic or politically motivated cooperative institutions.[5]

In 1978 the Rowntree Social Services Trust provided an initial grant for a body with a rather different philosophy, Job Ownership Ltd (JOL), under the chairmanship of Jo Grimond, the former leader of the Liberal Party. Though at pains to avoid a party label, JOL's philosophies are liberal rather than socialist. Its theme is 'enlightened self-interest' which it believes can make a cooperative more efficient than a privately owned or joint stock company. It believes in a substantial personal financial commitment by each cooperator in order to give the necessary incentive – normally a minimum of £2000. Since the Provident Societies Act forbids the personal ownership of more that 10 000 shares most JOL cooperatives are in fact incorporated under the Companies Act and often seek government funds under the Industrial Common Ownership Act of 1976.

JOL took as a model the Mondragon system (see below) and aimed mainly to persuade existing businesses to convert to cooperatives. It set out a number of possible examples:

(a) A family business whose owners are nearing retirement and have no natural successors able or willing to take over.
(b) 'Demerging' of a subdivision of a large company or conglomerate.
(c) A large company in an industry where changing markets or technologies make a split into smaller units an attractive alternative.
(d) An enterprise with a sound product but other seemingly insoluble difficulties such as bad management or chronic industrial conflict.
(e) A small business (probably mainly 'white collar') in which there is a strong desire to form a cooperative amongst the work force.
(f) A company which wishes to make itself invulnerable to the threat of take-over.[6]

The example of Baxi Heating described in Chapter 12 contains elements of (a) and of (f), though in that example the solution chosen was a partnership on John Lewis lines rather than a cooperative.

There are many many more organizations dedicated to fostering small cooperatives with varying degrees of political commitment and success such as the Commonwealth Development Fund (another child of Ernest Bader) which makes loans of up to £5000, normally for a year, the Mutual Aid Centre, local Cooperative Development Groups, the Industrial and Commercial Finance Corporation, the Centre for Alternative Industrial and Technological systems and the Socialist Environment and Resources Association.[7] It is perhaps stating the obvious to say that, while the commitment and dedication to work a 14-hour day can achieve miracles, a small cooperative seldom in practice achieves them unless it solves the problems of raising capital, maintaining cash flow and marketing – i.e. of sound commercial management.

SCOTT BADER

One of the largest and most successful of cooperatives is Scott Bader, founded in 1951 when, as stated earlier, the founder of ICOM handed over his successful chemical firm to its workers. Though large for a cooperative it is still a small concern – of about 350 workers. It has a strong idealistic and political motivation (Christian Socialism) but a major factor in its success has been that it has always had massive capital reserves. These were initially inherited from Ernest Bader's private firm and thereafter by successful capital formation. Its initial reserves in 1951 were £129 750 (including £50 000 in share capital). By 1973 its reserves and retained profits amounted to £1 187 194 and by 1978 to £3 396 989. It has been unnecessary to raise capital from the workers themselves who therefore have no personal financial stake in the business. Over the same period (1973–8) its reliance on loans decreased from 22.4 per cent of share capital to 4.5 per cent and its actual bank loans from £424 750 to £154 750. Its net return on capital has varied between 21 per cent and 34 per cent.[8]

Like the John Lewis Partnership, Scott Bader's democratic constitution is complex, with safeguards which have ensured that the democratic control has not hampered sound commercial decisions. When Ernest Bader retired in 1966 he left his son as Chairman of the Board for life.[9] Like Spedan Lewis and the Sieff's of Marks and Spencer, he has predictably been sneered at as paternalistic but all three firms have combined a contented workforce with commercial success and these, with good economic management, do seem to go together.

MONDRAGON

In contrast with Scott Bader and many of the other cooperatives described in this chapter the workers in the famous Mondragon cooperatives in Spain *do* subscribe capital, a minimum of £1000 of their own money, though this normally represents only about 20 per cent of the capital of an individual cooperative. Most of the capital, and the advice on organization and management skills, comes from the central bank which really controls the whole Mondragon complex.

Mondragon is a small town in the Basque region of Spain. In the surrounding area – about the size of Devon – there are 58 industrial and 18 other cooperatives employing a total of some 15 000 people.[10] Nearly half of the 58 industrial cooperatives have under 100 members, 14 have 100–250, 11 have 250–500 and six have 500–1000. Only one has more than 1000 and that is Ulgor, employing some 3500.[11]

The Mondragon experiment began with Ulgor in 1956, when Jose Maria Arizmendi, a priest, set up a small cooperative of two dozen workers. Ulgor has now become Spain's leading manufacturer of refrigerators, cookers and washing machines. In addition to its own 3500 employees it has hived off another 3000 in small cooperatives when they have become able to stand on their own. Ulgor's growth has been phenomenal.

Equally phenomenal has been the growth of the cooperative bank, the Caja Laboral Popular. This grew from a small community bank formed to help finance Ulgor and a small technical college. It now has over 70 branches and over 200 000 members and in 1977 made a profit of £4 million.[12] It has a democratic structure and is in effect a cooperative itself.

When a new cooperative is formed it signs a contract of association with the Caja. The Caja normally puts up 60 per cent of the capital and therefore has a controlling interest. A further 20 per cent is issued on loan from a special fund for this purpose and the remaining 20 per cent by the work force. In the case of a new cooperative, each cooperator has to put up about £2000, or about half of that if he joins an existing cooperative. If they leave, they forfeit 20 per cent of their stake and also forfeit any sum which has accrued to their account from the annual profit. Though somewhat ruthless this system has two advantages:

(a) There is very little turnover of members.
(b) The requirement for a financial committment keeps away the

footloose worker who drifts from job to job. Two thirds of the entrants come from steady jobs elsewhere.[13]

The democratic structure in an individual cooperative is normally as follows. The workforce elects a Supervisory Board by universal suffrage. This Board hires and fires members and appoints Directors and Executives to run the business. Amongst the Mondragon rules is that the ratio between maximum and minimum wages in any cooperative is 3 to 1 (or in some cases 4 to 1). This might be expected to be a disincentive to high quality managers but in practice this is not so because the Basque region is so much more prosperous than the rest of Spain that the managers' salaries, even with this restriction, compare very favourably with what they could earn elsewhere.[14]

The significant factors in the success of Mondragon are: sound economic management with a strong financial reserve in the bank; a personal stake by every worker in the business which would result in his facing considerable losses either if the business failed to prosper or if he decided to leave; and a democratic structure which tempers workers' control with safeguards to prevent managerial decisions being hampered. As a further guarantee of this last point, the bank has a direct involvement in management and its 60 per cent share enables it, in the last resort, to ensure that such decisions are commercially sensible.

The experiment, in addition to its evident social benefits, shows the way in which a bank can make money for its members (including the cooperators) by financing and overseeing cooperatives and provides a model which should be studied with a view to harnessing bank capital to partnerships and cooperatives in Britain, as discussed in Chapter 12. The Caja Laboral Popular, however, was originally formed for this purpose and grew from its own success. An established merchant bank might feel reluctant to take the risk and think it safer to invest its money in more conventional ways. Though it might be tempting for governments to invest in this way, the next section of this chapter suggests that there are very real snags to that, especially in Britain.

KME

Shortly after Harold Wilson's Third Labour Government was elected in February 1974 the Secretary of State for Industry, Tony Benn, played a leading part in forming three cooperatives with government capital – Meriden Motorcycles, the *Scottish Daily News* and Kirkby

Manufacturing and Engineering Company Ltd. (KME). All three failed. Meriden lasted longest, the Conservative Government writing off its debts up to £11 m in 1982, but it too finally went into liquidation in 1983.

The KME cooperative was the fourth attempt to rescue a previously successful subsidiary of BMC, Fisher Bendix, which had been manufacturing a wide range of consumer durables in the new town of Kirkby, Liverpool in the middle 1960s. After takeover in turn by Parkinson Cowan and then Thorn Electric without success, there was a third and rather unusual rescue attempt by International Property Development (IPD) in 1972, in which Harold Wilson, then MP for the Huyton Division of Liverpool, had played a prominent personal part in conjunction with Fisher Bendix's AUEW Convenor, Jack Spriggs. By early 1974, however, just after Wilson had won the February General Election and became Prime Minister, the factory was tottering towards insolvency and a receiver was about to be called in. Wilson's Secretary of State for Industry, Tony Benn, refused a request by IPD for a £1 m loan but invited Spriggs and his fellow convenors to a meeting in London on 18 June 1972 at which he floated the idea of a cooperative. By 11 July, however, the firm's credit had collapsed and Barclays Bank put it into receiversip.

Soon after the receiver, Roger Cork, had moved in and investigated the figures he announced that the work in hand would only employ 450 of the 1100 workforce. When he rejected the shop stewards demand that 1100 must be employed, they ejected him without allowing him to address the mass meeting of workers who thereupon welded the gates shut. Not surprisingly Cork failed to find a buyer. On 2 September Benn called a meeting which included Cork, Spriggs and the TGWU Convenor Dick Jenkins and told the convenors that, if they could produce sensible plans for a workers' cooperative he would give it full consideration.[15] Meanwhile the Government would temporarily provide up to £27 500 per week to support the receiver – presumably to enable him to continue to pay wages.

The convenors had already done some thinking following Benn's hint in June and were in possession of a constructive report prepared by a firm of management consultants, Inbucon, for IPD. This suggested that if the factory carried on with 1100 employees it could lose up to £2 million per year but that it could be made viable if it reduced the work force to 590, cutting out some of its unprofitable business (such as soft drinks) and concentrating on engineering work such as radiators, night storage heaters and contract press work.

On 13 September, Spriggs and Jenkins submitted an application for a grant of £3 901 000 under section 7 of the 1972 Industry Act, with which to buy on behalf of the workers the whole business except for the land and the factory which they would rent from IPD. Though they incorporated many of Inbucon's ideas, however, they preferred to keep the drinks business and to employ 913 people instead of the recommended 590.

They coined the name KME Ltd and proposed that the share capital be held in trust for the workers. The Board of KME Ltd would comprise 6 worker directors elected by the workers. This Board would appoint a Management Board comprising 3 of the Board's Worker Directors, with an outside chairman, 2 other executive managers and 4 senior managers to do the day to day running of the company. In the event, only Spriggs and Jenkins became directors on the KME Board, their real power being derived from their position as convenors elected by the shop stewards committee. They were thus at the head of two structures – the management and the trade union.

A cooperative executive committee of 10 was to be elected by the workers, but it was later reduced to 7 (i.e. 6 workers, one from each of the 6 trade unions, and one representative from management) and renamed the 'Cooperative Council'. The Council was, in fact, resented by the shop stewards, who felt that it was a threat to their own power. Within a year the Council had collapsed, leaving Spriggs and Jenkins at the pinnacle of every source of power – the Board, the shop stewards committee, the management board and, now, the cooperative of which Spriggs had become chairman.

The main weakness, however, lay in management. The Board appointed the former production director of IPD, Bob Lewis, as General Manager of KME. He was the only senior manager. He had a finance controller and a works manager but no marketing or commercial manager. Lewis himself lacked marketing or commercial experience and, since he was not a member of the KME Board, had little power of decision.

To make matters worse, the workers themselves were not really committed to the cooperative, which had not been their idea. They still regarded the role of their stewards and convenors as to preserve their jobs and ensure that they got high wages. As one of them remarked: 'I don't want stewards wearing two hats – I want them to fight the boss.'[16] Since the government had put up the money they probably regarded the Government as the boss, from whom the stewards could no doubt screw more money if required. They took little interest in the Council during its brief existence and showed little or no commitment to make the cooperative pay its way. In other words, they never regarded themselves as owners.

Spriggs himself did not help by announcing at the start that no one would be sacked, which predictably resulted in poor discipline. A 'score' system of payment was also introduced, whereby a worker was free to leave when he had completed his own daily score. Some did this by 10 a.m.

The result overall was commercially disastrous. At the start KME was losing £25 000 per week and this did not decrease – it lost £345 000 in the first 3 months and £750 000 in the first 6 months.

In 1976, KME applied for a fresh loan. Eric Varley had replaced Benn as Secretary of State for Industry and the application was refused. The company continued to ignore advice on how to improve sales and its own financial control and marketing were weak or non-existent. Various attempts to persuade other firms to mount a rescue operation failed and KME went into liquidation on 27 March 1979. Next day the Labour Government (with James Callaghan by then as Prime Minister) lost a confidence vote in Parliament and a General Election was called on 3 May in which the Conservatives were returned to power. KME had been born and died with the 1974–9 Labour Government.

Why did KME fail? Above all because it lacked the most indispensible ingredient of all for successful cooperatives – good commercial management. Its failure also underlined the impractability of dual control of management and trade union structures. Spriggs commented: 'There is no dual control. This company is the property of the unions.'[17] The experience suggests that this is even more impracticable.

As Patrick Wintour commented:

> A perhaps unkind inference is that Jack Spriggs and Dick Jenkins wanted all the power but little of the responsibility . . . and refused to appoint a high-calibre manager for fear he or she might be a rival influence. But, having centralized power, they could not bring themselves to take on the necessary obligation of shutting unprofitable sections or of sacking some of the workers in order to cut costs. Spriggs and Jenkins would have let the government wield the axe but they did not want the blood on their own hands.[18]

One inescapable conclusion is that the government is not the best sponsor for a cooperative. In this case, the government went further, and themselves took the initiative. Spriggs and Jenkins had not gone to see Benn to propose a cooperative but as convenors fighting for their members' jobs. When the only alternative to closure was a government funded cooperative they acccepted it but the workers were never committed to the idea and assumed that the government would not let it collapse – so the government became the boss to be milked.

CONCLUSIONS

Consumer cooperatives like Britain's 'Co-op' are not workers' cooperatives and in practice differ very little from joint stock companies. The two most successful cooperatives or groups of cooperatives considered in this chapter are Scott Bader and Mondragon. Both have the advantage of strong financial reserves and constitutions which, like that of the John Lewis Partnership, do ensure that the management is strong and not hampered in taking sound commercial decisions. Scott Bader has no workers' capital, having started from the base of a successful company in full operation and a philanthropic owner. Thereafter, however, it has largely generated its own capital by successful trading and incorporation of its profit. A rising turnover reaching £22 million with 350 workers is a remarkable achievement.

The Mondragon group is much larger. Its success is built on the impressive growth of the bank which owns 60 per cent of the capital of the constituent cooperatives and provided management expertise for them. The bank also maintains its reserves by providing a thriving local banking service to the Basque community in the surrounding towns and villages. One of its cooperatives, Ulgor, employs 3500 people but none of the other 75 have more than 1000, and most under 100.

Of the other bodies which sponsor or fund small cooperatives, ICOM and ICOF (both from the Bader stable) have been the most successful, though, in contrast with the Mondragon cooperatives, the workers have no personal stake beyond a £1 token share. JOL has tried to follow the Mondragon example but thus far has produced few cooperatives. The Government CDA and the Cooperative Bank have created many more.

KME, the subject of the final case study, seems to teach almost every lesson in how not to launch and organize a cooperative and was not so much a workers' cooperative as a form of nationalization with management vested in the shop stewards committee. The result, despite good intentions, was disastrous.

The total number of people in workers' cooperatives in Britain has been growing fast. In 1982 there were 6000, growing to 9000 by the summer of 1983, the number of cooperatives increasing over the same period from 500 to 900. As these figures indicate, their average size is small. The failure rate is about 10 per cent.[19] With no 'Mondragon' in Britain and the 'Co-op' not being a workers' cooperative, the total of all put together is a fraction of the workforce of the John Lewis Partnership (Chapter 12). Yet the Scott Bader and Mondragon experiences do suggest that, given sound finance with strong unhampered management and a work force committed to commercial success, they can prosper and grow, and provide a satisfying work environment so the all party support for them is fully justified.

14[1] Marks and Spencer

A MODERN BRITISH FAIRY TALE

Various forms of partnership, self-management and cooperatives were described in previous chapters. In the next chapters a number of more conventional joint stock companies will be examined, focusing on three in which participation and communication between management and the work force at all levels seem to work well. One retail and a number of manufacturing firms will be included.

One of the largest and most successful retail chains in the country is Marks and Spencer and it is no coincidence that, as well as keen commercial efficiency, it has always enjoyed a high reputation for communication and staff management. The story of Marks and Spencer is a modern British fairy tale. In 1884 Michael Marks, an almost penniless Jewish refugee from Russian Poland, was given £5 credit by a kindly wholesaler (Issac Dewhirst, whose firm remains one of the principle suppliers of M & S to this day) with which he bought haberdashery to peddle around West Yorkshire and from this small start he quickly developed his famous chain of Penny Bazaars. In 1894 he formed a partnership with Dewhirst's accountant, Tom Spencer, from which has emerged Britain's fifth largest company in terms of capitalization, employing 48 500 people, with group sales of £2505 million in 1982–83. It has become the largest clothing retailer and the fifth largest food retailer in Britain less than 100 years after Michael Marks borrowed £5 to start his business.

MANAGEMENT AND GROWTH

Most customers are impressed by the high morale of M & S sales staff. Despite having a predominantly female – and married – work force, stability is high. Of the 48 500 employees, 10 000 have been with the company for more than 10 years; 20 000 more than 5 years; and 97 per cent of the full time and 89 per cent of the part time employees more than one year. Good management, good communication and exceptional welfare benefits play a big part in this but the foundation of high morale is, as with John Lewis, sustained growth and commercial success.

For, despite the recession, the business has continued to grow. Currently about £100 million is being spent each year in building new stores or extending existing ones. The firm enforces stringent quality control on its suppliers and, since M & S contracts are too good to lose, the suppliers do achieve these standards so all the staff involved have a pride in what they sell. The policy is not to pioneer high fashion but to organize production to provide and sell briskly what the public seems to want most at that time. This, too is stimulating and satisfying for those doing the merchandising and the selling.

In any field of endeavour people are happiest if they feel that their work is well organized. M & S management is efficient and personal and this reflects the system of recruitment and training of managers. Competition for entry is intense. There are, for example, about 100 applications for each graduate trainee appointment, but in 1983 the firm was 30 short of filling all its vacancies to the standard required. The final selections are made in a two day residential 'houseparty' selection board. There are, however, other means of entry to management. There are recruitment schemes for school leavers with 2 'A' levels, for people with Higher National Diplomas or over 21 with suitable work experience, and there is promotion through supervisory and management grades.

Basic management training takes 13 weeks including 6 weeks of course work. Thereafter, managers are pulled out for training and development courses throughout their careers. Managers are regularly moved so that they have a wide span of experience – 3 or 4 years is normal tenure in each appointment.

As with John Lewis, trade unions are given facilities for convening meetings to recruit members but they get little response mainly because of the predominance of female workers and the general confidence in the existing channels for participation and communication.

At working level in a store, an Assistant Staff Manager is normally responsible for 50 to 60 staff, working through one or more levels to a Staff Manager. The Staff Manager works to a Divisional Personnel Manager (there are 12 divisions each controlling about 20 of the 262 stores) who in turn answer to a Personnel Executive at Head Office and thence to the Director of Personnel on the Main Board. This structure is parallel with but independent of the commercial management structure. In a medium sized store, for example, this starts with a supervisor managing 10 or 12 sales staff and goes up through departmental managers to the store manager. The Staff Manager in a store is an integral part of the management team but, having direct access to the divisional office, can be seen by staff as representing their interest, through to the Head Office if necessary.

One of the strengths of M & S is that managers do have clear responsibilities and are judged by their performance. The sales of each store and each department are made available to the Board and senior management each week with the percentage change over the previous or equivalent period. There is an old army dictum that 'the person who writes your Annual Report is your real boss'. This goes for M & S, in each sector of Commercial, Office and Personnel Management. Appraisal Forms are detailed and comprehensive and are seen by, read by and initialled by the 'victim' – again as in the army. Thus every manager and in fact every employee knows where he really stands in the company's estimation and records, so he can assess his own prospects.

COMMUNICATION

One problem which plagues any company growing to giant status is that management can become remote. The common answer is decentralization. Lord Weinstock is reputed to have said that, when a unit became bigger than 200, he opened a new factory. M & S, considering whether to increase the size of their 'flagship' store at Marble Arch, (where the full time and part time staff is now in total about 1400), based their decision on whether they could find a way to subdivide it into manageable entities. The desirable maximum is 500 but, like Lord Weinstock, M & S regard 200 as an ideal, in which the store and staff managers can know everyone personally and know something of their domestic and personal problems. This can be achieved on a floor or departmental basis but that does not solve the problem of remoteness of the top management. It is in this field of management communications that M & S can teach some of its most valuable lessons.

M & S attempt to solve the problem by top management going to the shop floor rather than by inviting elected shop floor representatives to sit on the Board. The firm is probably too big for a successful system of elected worker-directors, since the great majority of those who elected them would not know them personally and they would soon become as remote as the remotest of the managers – as has been found in a number of large companies and nationalized concerns which have worker-directors.

In M & S, each Director on the Board will visit a substantial number of stores during the course of the year and, whenever he visits a supplier, he will normally try to visit the local store as well. In addition, each Director including the Chairman makes a point of visiting his own local store virtually every week. By getting to know a specific store in depth, the Director can constantly relive his own junior and middle management experience and so avoid losing touch with the changing conditions of trade at the sharp end.

The Chairman himself travels incessantly. He also follows a routine of telephoning the managers of 10 to 12 of his 262 stores every Friday and getting them to update him on their store, trading developments and events in their town. These regular conversations benefit both the Chairman and the managers, and, of course, the company as a whole.

Throughout their history M & S have always had excellent communication downwards but in 1970s, with their continuous growth, they felt the need for better communications upwards. They therefore introduced, in addition to the existing communications structure through staff managers, a new concept of 'communication groups'.

The communications groups were introduced in 1975 because the Head Office wanted a more formal channel of communication which would give the views of all categories of staff. These communication groups vary in size from 5 or 6 in a small store to 15 to 20 in a large one. In the very large stores they may operate in two tiers; three groups, for example, each representing a floor or department of manageable size, meet first; every group then sends 5 or 6 representatives to the meeting of the group representing the whole store.

The method of election varies, since stores vary in size from 30 people to over 1000. The staff themselves decide whether to use a ballot or some less formal method. Candidates must have served at least two years with the firm to be eligible. They are elected for one year and may be reelected for a second year after which they must stand down.

The group elects its own Chairman who holds office for one year. It meets as often as it likes and at least every two months. The meetings start with no management present. The group decides what subjects it wants to discuss and then asks the Store Manager and Staff Manager to join it. Managers answer all the queries they can and the groups are free to preserve the anonymity of the source of the query if they wish in order to encourage frank criticism. If the manager is unable to answer a question or if it relates to general policy the Chairman of the group will write to the appropriate person in Head Office. The proceedings are minuted and copies of the minutes sent to Divisional and Head

Offices. Managers are given short (2½ day) courses in group dynamics as part of their preparatory training.

In common with other participatory structures, these groups have settled to their task and have proved to be constructive and responsible. They are now regarded as an indispensable part of the running of the business. They are kept informed of company results, problems, progress and plans and of external problems affecting trade. They are used by management as consultative bodies and they are asked to conduct research projects or to seek views on local issues.

There are two house journals: *Sparks*, published four times per year, contains normal magazine type articles; *St Michael News*, published eight times per year, is used to inform staff of forthcoming developments so that they can start thinking about the future. It is a colourful and immaculate production with pictures of the standard to be expected in a promotional brochure and is used to alert staff to future marketing opportunities for, say, new lines of clothing which are being tried out successfully in a pilot scheme in about 10 stores and the decision has been taken to market them nationwide.

Another communication technique, which also serves as a training function, is worthy of mention. One day each week the store opens for customers 30 minutes later than on other days and these 30 minutes are used by management for briefing and discussion with the staff. The time is often used for training, sometimes by bringing in outside speakers on, for example, hygiene or the role of the police.

M & S have also experimented with the 'Quality Circle' system as practised in Japan (having originated in Britain), whereby small groups are trained in the techniques of identifying problems and their causes and in arriving at solutions which are then communicated in the form of a presentation to senior management. This technique probably has more application to manufacturing than to retailing and M & S found that the function was already fulfilled by existing management communications and the communications groups, so the Quality Circles as such gradually declined into disuse. Experiments in this communications field, however, are continuous and at the time of writing a 'quality circle' comprising representatives from each department in one store (e.g. sales floor, stock room, staffing, administration) was meeting on a peer group level to discuss ways of improving performance as a whole.

There is an active suggestion scheme. Suggestion forms are available on a 'help-yourself' basis and the suggestion goes into the suggestion box without reference to the local manager, though he will certainly see it if it is accepted. Some 2000–2500 suggestions are received each year and 7–10 per cent are accepted. For those accepted there is a minimum award of £15 but it is often very much higher; one warehouseman, for example, received £500 in 1983 for the design of a new method of displaying rugs which was later adopted in 50 stores. There is no upper limit.

BENEFITS

During the depression in the 1930s the Vice Chairman, Israel Sieff, was walking round a store and came upon a sales girl who looked on the point of collapse. He asked her if she felt well and she anxiously assured him that she did. The Store Manager was instructed to find out about her and it transpired that she had two brothers and that both they and her parents were unemployed. Apart from the dole, then a pittance, all they had to live on was what she earned so they all went short, and she was desperately afraid that, if she were seen to be ailing, she would lose her job.

This incident convinced Sieff and The Chairman, Simon Marks (his brother-in-law) that they should ensure that no-one working for M & S must be hungry and they introduced what was then a revolutionary idea – a staff canteen in every store, financed by the company. That their compassion was genuine is beyond question but there is equally no doubt that it made commercial sense to have a well-fed staff.

The practice continues today and in 1983 every M & S employee received a three course lunch, plus coffee and tea breaks, for a token sum of about £6 a month. This works out at about 30 pence per working day, or it can be taken on a daily basis.

These subsidized meals cost the company £11 million pounds per year. To this must be added the non-contributory pension scheme (£27.5 million), medical service (£2.6 million), staff discount (£1.9 million) and sport and other welfare funds (£0.8 million) to make total benefits in kind costing £43.8 million per year – which is £903 for each employee full or part time.

Employees receive a Christmas bonus of about 10 per cent of salary. In addition, every employee with more than 5 years service qualifies for the profit-sharing scheme averaging 4 per cent of his or her earnings in the form of M & S shares paid for as a charge on profits and added to the company's capitalization. These shares are theirs to keep or sell as they wish, but the greater majority retain and build up their holding.

These shares currently cost the company about £4.6 million per year (which is not included in the £43.8 million benefits in kind).

Though not as expensive as the catering subsidy, the medical service (£2.6 million) is again far beyond anything normally found. At the Head Office (which, including all the 1500 buying staff, houses 3500) there are 6 full time doctors with dentist, chiropodists, physiotherapists and other staff. There are in addition 240 local doctors retained for regular part-time work in the stores, varying with size from once a month to two or three times a week. It is primarily an Occupational Health Service, researching and preventing health hazards (e.g. those arising from long periods of standing or repeated bending to pick up heavy loads), directed to physical and mental well being. The return on this £2.6 million is indirect – contributing to very low absenteeism from sickness and all other causes (3 per cent), low staff turnover and the general contentment with 'a good place to work'. The result, again, is reflected in the bottom line of the balance sheet.

Put another way, M & S add, in free shares and benefits in kind, between a quarter and a third of each employee's salary, equivalent to nearly half as much as all other running costs, and not far short of the total they pay out in dividends (£67.1 million in 1983) to shareholders. Few firms could produce figures comparable with these, and they are convincing proof that the company does regard the people working for it as its most valuable asset, which, in a retail firm, they undoubtedly are.

To sum up, Marks and Spencer, like John Lewis, enjoy many advantages which make their problems easier than those of a manufacturing concern. Working conditions generally are more agreeable and more varied than in most factories. The company, with its predominantly female and flexibly employed work force, coupled with constant expansion, has thus avoided the redundancy problems from which so many other problems flow. This also leads the work force to prefer not to join trade unions so, although the company does not have this particular upward channel of communication, it has caused them to develop and rely more on their own channels which are thus more direct and personal. Another advantage common to John Lewis and M & S is that it is easy to fix clear responsibility on managers of stores of a manageable size. Thus good management, good communications and a remarkably generous range of benefits in kind have provided the profits and the high staff morale to enable them to maintain the commercial success and the growth, which both avert the problems and cover the cost of the good management and the benefits.

How far their example can be applied to other industries, especially manufacturing industries, is the subject of the next chapters.

15[1] Alvis

BACKGROUND

Alvis was founded in 1919 and has concentrated on the upper end of the market, in its early years on sports and racing cars and more recently on military vehicles. Its individual workers, unlike those in volume car production, have always been able to identify their work with particular vehicles or engines, giving them some job satisfaction. It is a medium sized firm of about 1850 employees, the great majority of whom work in a single factory, and its problems and their solutions can fairly be compared with those of Westfalia Separator, the West German firm described in Chapter 4.

By 1930 Alvis had an established reputation in motor racing and amongst rally enthusiasts and extended its activities to aero engines including the Merlin in collaboration with Rolls Royce. After the Second World War the firm concentrated almost exclusively on military vehicles, in which it had built up an expertise during the war. In 1965 it merged with Rover and experienced its first labour troubles. This merger led in due course to a somewhat unhappy incorporation into British Leyland which ended in 1981. Since 1967, Alvis has moved into tracked vehicles, the most famous being the Scorpion light tank, first produced in the 1970s and used with great success in the war in the Falkland Islands in 1982.

This gave the firm a stable order book, based on a 7-year contract with the Ministry of Defence (MOD). This period (the 1970s) coincided with one of trade union militancy arising from the 1971 Industrial Relations Act. The stable order book enabled the Alvis work force to go to the offensive and they became the best paid engineering workers in Coventry (averaging £130 in 1974–75 at a time when the average manual wage was about £50). This was achieved, however, at the price of repeated strikes which caused the firm to fall behind on contracts. As a direct result of their high labour costs and unreliable delivery the firm's reputation with MOD was eroded and Alvis lost (to GKN) a lucrative long-term contract to produce the MCV 80 in 1975–76. Their other MOD contracts ran out at the end of 1981. The firm was sold by BL in 1981 to United Scientific Holdings.

This history and its consequences in the recession of 1979–82 created the climate for the positive and thus far successful effort to improve industrial relations by better communication and participation. The long term tradition of identification with the product has survived. The firm only produces 8 to 10 vehicles per week and each normally takes 12 months to manufacture from the first machining of parts, through sub-assembly and assembly to the final testing of the assembled vehicle. Wages, though they have not kept pace with inflation since the golden years of 1974–75, remain high (£4 per hour or a factory average of £152 for a 39 hour week in 1981 compared with a national manual average of about £100) but the end of the MOD contract means that the firm has to fight constantly for its share of the market. Since this must now largely be by exports to a world itself in the grip of recession, there is intense competition not only from other British firms but also from France, Germany and Brazil. The work force have few illusions now and this has weakened the power of the unions. There have been no significant strikes since that of 1979, which was a national strike called by the AUEW, not a dispute with the firm. Many of the older workers and shop stewards now give the impression that they are aware of the price they have paid for using strikes to drive hard bargains in the 1970s – the loss both of a stable order book and of trade union bargaining power.

CURRENT PROBLEMS

Apart from the order book and the recession, the main industrial relations problems at Alvis lie in the necessity to automate to survive, and in getting agreement on the pace and implementation of the change. Some union officials have an ambivalent attitude to this problem: on the one hand they draw unfavourable comparisons with productivity in Germany and Japan, blaming it on inadequate post-war investment by British industry; on the other hand they not unnaturally fight the redundancies which such investment, albeit belated, will bring.

Computerized Numerical Control (CNC) was introduced in 1982. This will mean slimming down the work force from 1850 to 1500 over the next few years. As the management has put it to the unions, it is better to save 1500 secure jobs than to go bankrupt and lose 1850. There is, obviously, concern in the trade unions that this process will continue with further job reductions over the years and some feel bitter that management seem not to share their alarm about this, or to care about what it means in human terms. Realists on both sides, however, know and admit that within their lifetime surviving engineering firms will be using robots for all repetitive processes and that the task of labour will be to manufacture, programme and service the robots. The problem of converting this industrial revolution, like previous industrial revolutions, into equal numbers of more satisfying and more rewarding jobs with more leisure was discussed in general terms in Chapter 1. The general will apply to the particular in Alvis as in many other engineering firms.

Another continuing problem is that of differentials, During the middle 1970s the process workers used their collective power to obtain generous bonus and incentive schemes as the price of their cooperation in meeting contracts. As a result, the supervisory grades, maintenance staff and others for whom no element of piece work was applicable lost the monetary advantage of their skills. The wage restraints of the 'social contract' period (1975–78) enabled some of the inequities to be ironed out, but adjustment for hourly paid workers proved easier than for salaried staff, so many anomalies remained. Thus in 1981 the weekly earnings of a welder and a technical foreman were within £2 of each other and skilled fitters and toolmakers earned more than design engineers with HNC or university degrees. The resolution of such problems almost invariably leads to inflation and higher costs per unit of production because, apart from reduction of overtime, no one falls back and every adjustment must therefore be a leapfrog for someone. The art lies in combining this process with the higher productivity which will arise from automation so that no one loses earnings in real terms. Alvis management believe that the technical capability already exists for a 30 per cent rise in productivity, and that this must be phased in gradually to maintain a steady rise in real wages without losing competitiveness in the evolving market.

Another problem arises from the universal desire of human beings for power – or at least the ability to maintain their own options. Management need to maintain flexibility, keep costs down and therefore to minimise wage leapfrogging; they may be wary of providing ammunition for any group of workers to achieve it at the expense of others and this can conflict with the desire to communicate as much as possible to shop stewards and workers to earn their cooperation in introducing new equipment and new techniques. The type of shop steward who wishes to use industrial conflict as a means to bring about political change would certainly make use of any such information for this purpose so provides the strongest deterrent against management communication. The more moderate shop steward, whose aims are purely to get the best he can for his members may also, however, make use of such information to achieve that end and also to strengthen his own support by projecting an image of one who will fight management all the way to get it. He dare not be seen as a management stooge or he will merely be supplanted by someone more militant.

THE STRUCTURE OF THE COMPANY

Of the 1850 labour force at Alvis in 1981, all worked at the main Holyhead Road factory in Coventry except about 130 at the vehicle test site at Bagington. There were 60 senior managers, 70 shop stewards representing 1300 manual workers and 20 staff representatives for the 450 white collar staff. The family tree was as follows (Figure 15.1):

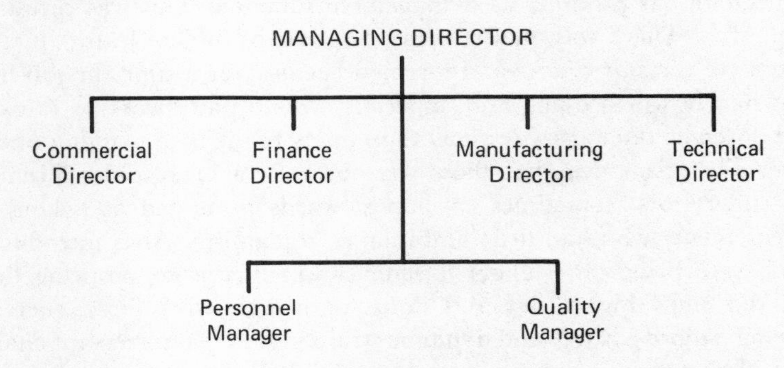

FIGURE 15.1 *Alvis family tree*

There were 10 Trade Unions in the company of which the largest was the AUEW and the Convenor was normally elected (by the 70 shop stewards) from that union. He headed a Negotiating Committee of 8 shop stewards representing the hourly paid manual workers and this Committee negotiated all agreements on wage rates, conditions, manning, redundancies etc. with the management. The Committee also tried to resolve inter union disputes, though some of these are between manual and staff unions – e.g. over who should programme the CNC machines.

The Convenor worked full time on trade union duties but was paid and provided with an office by the company. Shop stewards had paid time off for one meeting per week (2–4 hours) and could get additional time off if needed (e.g. during a dispute) by asking their foreman.

COMMUNICATION AND PARTICIPATION

The official channel for communicating management information and decisions is through supervisors and foremen to the shop floor. As the unions extended their power after 1945, many shop stewards in British industry usurped this power by the threat of bringing work to a halt in their shop (often with unofficial strikes) if any instructions were passed to the work force other than through them. This seriously eroded the authority and status of supervisors and foremen and was inimical to high productivity, accounting perhaps more than any other factor for Britains's very poor record of productivity compared with those of Germany and the USA.

This unhealthy situation has largely been remedied and the management chain has resumed its management function. This was certainly so in Alvis. There was, nevertheless, a residual problem in the quality of some of the supervisors and foremen because for a time the job had been hardly worth doing and, especially where piece work rates were paid, foremen often drew less pay than many of the men working under them. The result was that there were in 1981 a lot of older foremen and supervisors (sometimes ex-shop stewards promoted as a kind of appeasement) who had little ambition or dynamism. Alvis introduced a policy to bring on younger foremen and supervisors, ensuring that they did enjoy higher pay and status than their work force, thereby infusing a more positive and dynamic strain into the management chain.

At the same time, however, shop stewards were always *informed* of management decisions, preferably some hours before they were communicated to the work force. They generally accepted that this separation from the executive channel (but not from the information channel) enabled them to carry out their negotiating function better without conflict of interest.

Apart from this formal channel, Alvis had two other regular means of communication: through the Consultative Committee and by direct communication.

The Consultative Committee met roughly once per month. It had 18 members, including the Managing Director, the Manufacturing Director, the Industrial Relations Manager, the Publicity Manager and senior representatives of all the unions. It was *not* a negotiating body. Its prime function was to communicate information about the company's performance, marketing and prospects and to invite discussion of these. After each Monthly Board Meeting, the company's full trading results and market prospects were passed to the Consultative Committee within 36 hours and discussed. The Managing Director and Manufacturing Director were open to any question on the company's activities, including, for example, sales, profitability, the strength of the order book, manning levels and levels of inventory. One of the problems at such meetings (as was brought out in the Dunlop case study) was to avoid spending time on trivial matters best left to one of the many other Committees which existed for that purpose – e.g. the Production Committee, Health and Safety Committee, Resources Committee etc. Attempts to keep such matters away from the Consultative Committee were not entirely successful – e.g. some typical minutes examined in 1981 recorded questions on trees, gear-box cases, social club, recruitment procedures and Alvis badges. This could be frustrating but was hard to cure, for questioning could not be called 'open' unless members were free to ask the questions they wanted to ask. The management did, however, do their utmost to foster questions about fundamental problems relating to the company's viability and prospects, not only because it was in everyone's interest that these were communicated to the work force, but also because realistic negotiation of wage rates, manning levels, retraining, etc. could only be done if both sides were aware of the facts.

The non-negotiable status of the Consultative Committee was, however, enforced as strictly as possible, though leading questions were inevitably asked. Once again, a balance had to be struck between an uncompromising negative, 'That is a matter for the Negotiating Committee', and an atmosphere of frankness in sharing common problems. The biggest asset in the end was a good personal relationship, on the basis that everyone in the room was earning his living with the company (there were no 'owners' present).

One of the uncertainties in such a system is how far, and how accurately, each individual representative will communicate the information to his members. A shop-steward who is endeavouring to carry his men with him in some on-going negotiation, perhaps in competition with another union or, in some cases, because he has a political motive will be tempted to release only selective information to his members, in the same way that he may accuse management of releasing only selective information to the stewards.

Partly to avoid such problems, Alvis management also offered direct briefings by senior managers or the Manufacturing Director himself to the entire work force in a particular shop, or to the complete night shift etc. These were, of course, expensive not only in management time but in loss of production time. In theory, a half-hour briefing of 300 men at £4 an hour cost £600, though in practice it was hoped to make this up in higher productivity and fewer stoppages.

This direct communication was very popular with the work force but was regarded with reserve by some convenors and shop stewards. The managers were naturally as frank as they could be, and this could sometimes pre-empt the work of the Negotiating Committee. Some stewards, however, clearly resented the erosion of their own power if their members received the full facts from management direct and believed them.

In 1982 Alvis management had not yet introduced Briefing Groups, as practised in Japanese companies and commended by the Industrial Society, though a similar idea was under consideration.

In all these meetings, especially in Direct Communication meetings, efforts were made to allay union suspicions and resentments by briefing union representatives before the meeting on what the Director or Manager was going to say and inviting them to comment at the meeting itself.

Nevertheless, union scepticism about both communication and participation persists. Many shop stewards regard true participation as a myth, claiming that it could only be participation without reponsibility. There are no Worker Directors at Alvis and the unions do not want them. If they were appointed by the unions, they would face a conflict of interest between their responsibility as Board members and as union representatives, thereby inhibiting proper collective bargaining. If, however, they were directly elected by secret ballot from the shop floor (as in Germany), the unions believe that this would not only erode their own power but the Worker Directors would themselves be discredited in the eyes of the work force, having joined 'the other side'. The German experience largely refutes this view, since Worker Directors who frankly support and explain the benefits of higher productivity, lower costs, higher sales and higher profitability for investment do in fact get reelected by their own work force by secret ballot. Nevertheless, because it does erode their power, unions can scarcely be expected to support the idea of directly elected Worker Directors: and some directors and managers, reluctant to share confidential commercial information with workers representatives for fear both of leaks to competitors and of prejudicing their own negotiating position, share the unions' reluctance to accept them. A great deal of patient building of personal trust will be needed in British industry before such participation can be successfully introduced. It may well be that self-management in one of its various forms may in the end be easier to attain as this leaves no room for doubt that the workers lose as much from bad commercial decisions by their elected directors as they can from good ones. To evolve either to genuine participation or to self-management, however, the first stage is to develop trust, for 'we' to replace 'them and us' and Alvis, having been severely hit by aggressive industrial conflict and wage inflation in the 1970s, have advanced a long way along this long road of building up such trust.

16 GEC Telecommunications

THE COMPANY AND THE UNIONS

The General Electric Company (GEC) provides an example of a giant manufacturing corporation, highly unionized, which is succeeding in adjusting to change in an environment of rapid technological advance involving inevitable closures of outdated factories and cutting, redeploying or retraining its work force as new and more automated techniques replace the old. The subsidiary company selected for this study, GEC Telecommunications, was chosen because it has shown how these problems can be handled without undue disruption through industrial disputes, above all by good management-labour communications.[1]

GEC is Britain's third largest corporation in terms of capitalization (after BP and Shell) and its largest in terms of labour force (200 000 worldwide, of whom 135 000 are in UK). Sales are about £2500 million, of which one third are exports. Its strength lies in its decentralisation, with 130 UK-based and over 100 overseas subsidiaries. Some of these are small companies taken over intact by GEC, which continue to operate with a great deal of autonomy under their original names (e.g. Ruston Gas Turbine, the Express Lift Company, Hotpoint). Others are subsidiaries created and made autonomous within GEC as part of its own growth.

GEC Telecommunications Ltd is one such autonomous subsidiary employing (in 1982) 18 500 people, of whom 8000 are female. The company is in turn split into four divisions spread over five areas of the Midlands, Wales, the North East and Scotland, with additional production facilities in Australia, Canada, New Zealand and Nigeria.

This study was focussed on Coventry where 10 000 of the 18 500 work, over 5000 of these in the Stoke factory where the GEC Telecommunications Company Headquarters is also based. The other 5000 are spread round four other Coventry sites, two having about 2000 each and the other two 500 each. To complicate matters the four Product

Divisions (Telephone Switching, Transmission, Private Systems and Telephone Divisions) are themselves spread across the five Coventry factories and some of the outlying ones. In general, the smaller the factory the better the industrial relations atmosphere, though cutting across this is the well established experience that women are less militant than men and white collar staff less militant than manual workers.

Management of industrial relations in GEC Telecommunicatons has to steer a course between a Scylla and Charybdis. Much of the work is tedious and intricate, but the only solution to this (and to competitive survival) lies in automation, and this involves both capital expenditure and redundancy. The tedium is alleviated by a generous bonus scheme (see below) but automation inevitably causes friction with the trade unions who, on the one hand, oppose redundancies and, on the other hand, drive as hard a wage-rate bargain as possible when new methods and machinery are introduced – bargains which, if driven too hard, merely make the new process uneconomical, cause sales to fall and make further redundancies inevitable. It is to the credit of both management and unions, and in particular to the mutual confidence engendered by good communications, that the company continues to be profitable with a low record of strikes. Other factors, however, are that the products are generally small so that large buffer stocks can be held; also that it is possible to second-source almost all the materials; so it is less easy to make a strike effective than in many heavier industries.

The 10 000 workers in Coventry are covered by six trade unions, about half being in the two main manual unions (AUEW and TGWU), the remainder being technicians and white collar staff in ASTMS, ACTSS, APEX and TASS. There are between 300 and 350 shop stewards (i.e. about one for each 30 workers).

There is no closed shop but the Company has signed Membership Agreements with the technical and white collar staff unions. Under these Agreements, if the union can demonstrate that 75 per cent of the staff in a specified area or department are members of that union, that area will be established as a 'membership unit'. Thereafter the management undertakes that any *new* eligible staff joining the company in that area will be required to join that union within a month or they will move them to a different area. Non-union workers already with the company may, however, be transferred into that area without the obligation to join that union, subject to the number who are so transferred not exceeding an agreed maximum (normally 5 per cent). Under the Agreement the company also undertakes to encourage eligible members to join the union on the grounds that 'it is in the interest of all concerned that these unions should be truly representative'.[2]

These Membership Agreements work well though the number of membership units is small (about 5 per cent of the areas in which they could theoretically be applicable).

In the hourly paid unions – AUEW and TGWU – formal agreements have proved unnecessary since virtually all hourly-paid workers are members of one of these unions by choice. There is, in practice, a greater divide between staff and manual unions than between the unions and the management, since each union's main concern is to protect its own area and membership.

THE BONUS SCHEME

Unlike most engineering companies in Coventry, the company does not use Measured Day Work (MDW) but pays by results. The repetitive nature of much of the work makes incentives essential and the unions accept this. There is a good deal of piecework, which is popular with the work force as it gives them the maximum option in how they work. The bonus scheme has already been mentioned. The bonus is calculated on the ratio of sales to wages and salaries paid monthly but computed as a rolling average over the previous 12 months to smooth out variations. It is computed for the company as a whole and, though this may seem to give little incentive to the individual, it does engender a realization of the interdependence of the company's activities. Thus, a dispute in another factory may cause a worker to lose part or all of his bonus through no fault of his own. During the national engineering dispute in 1979, GEC workers received no bonus at all in one month and they also lost a lot from the 1980 steel strike. The scheme is certainly effective in making people who kick through their own goal very unpopular with the others and it has worked well over the five years in which it has been in operation. The average bonus in 1982 was about 7½ per cent of the basic pay scale.

PARTICIPATION

Formal participation agreements are mainly confined to areas not directly concerned with production such as Canteen and Joint Health and Safety Committees, in which issues are rarely polarised between unions and management but more often between the interests of different departments or groups of workers. On the production side the unions themselves are wary of having worker-directors involved in management because, if they represented the unions, their position would be anomalous whereas if they were elected directly by secret ballot by universal suffrage on the shop floor they would erode the union's authority. In either case, union officials and shop stewards feel that worker directors would not have enough knowledge to do themselves justice on the management boards, and would soon be regarded by the workers as 'them' rather than 'us'.[3]

The company therefore bases its industrial relations on three other pillars:

(a) Good management achieved by good selection and training
(b) Good working and personal relations between managers and union representatives at every level down to charge hands and shop stewards
(c) Good communication upwards and downwards both through management and union channels.

MANAGEMENT SELECTION AND TRAINING

As in most other firms, and in the army and the police, there are two ways of entering the management ladder: from the shop floor or as a graduate. The shop floor route, through charge hand, foreman and supervisor, provides a valuable cross fertilization between management and unions in that shop stewards often become supervisors and this leads to better understanding. At the same time, in a company with a large proportion of technical and salaried staff, a large number of junior managers are themselves members of the white collar unions.

The recruitment of graduate managers provides a dilemma (as in most similar industries) in that the company needs to attract the best graduates if it is to have able, young and dynamic, middle and higher managers in the future, and these graduates will be attracted elsewhere unless they see reasonable prospects of quick promotion and responsibility. On the other hand, to retain the respect of older supervisors and shop stewards, the management trainees must spend long enough on the shop floor to be seen to know and understand the job at working level. These factors inevitably conflict and demand a delicate balance to be struck. The priority, however, lies in developing them into good professional managers because, in the end, respect earned for efficient and sympathetic management closely involved with the shop floor will override considerations of age and background.

With this in mind, while the company does take a number of management trainees after graduation it prefers to recruit them at 18 and to support them on a five years 'thick sandwich' course. This comprises one year's practical experience on the shop floor, then three years at university followed by the fifth year on the shop floor again. (This is preferred to the three year 'thin sandwich' course offered by some universities and polytechnics in which the student alternates between six months at the university and six months on the shop floor.) The company gives the trainee an incentive by topping up the local authority student grant while he or she is at university.

During the two years on the shop floor (or in the drawing office or commercial departments, etc.) the graduate trainees may become foremen if they show that they have the quality and the confidence of the work force. Thereafter if they are good, there are avenues for fast stream advancement, coupled with periodic courses for middle and senior managers at each stage of promotion. There is also an annual assessment which is discussed with the immediate superior manager – again a normal practice but dependent on the frankness with which this discussion is conducted.

To encourage good working relations, the company runs regular joint courses at a country hotel, far removed from shop-floor pressures and from rival attractions so that members spend their evenings there together. There are equal numbers of trade union members and management, including shop stewards and foremen as well as more senior managers. They solve problems and carry out simulation exercises in mixed groups, often playing roles other than their own. They become friends and learn to see their future problems from all sides.

The heart of good working relations, however, must always lie in management at all levels being seen constantly on the shop floor, being seen to understand the work and regularly discussing it and seeking the views of the people on the job. The company does see to it that its managers do this and GEC's decentralized structure makes it possible. Though in some other companies such dialogue is resented by trade union representatives, relations are such that it seems generally to be welcomed in GEC.

COMMUNICATION

Supplementing this day to day shop floor dialogue, the company devotes a lot of management time to more organized communication, in the knowledge that this time is far less than managers would spend on industrial disputes if such communication were lacking.

While trade union officials and shop stewards are kept informed, care is taken to ensure that information reaches the shop floor workers first through the management chain – i.e. from their own foreman. If they were to hear it first from their shop steward the foreman's authority would be eroded. At the same time, information must flow up as well as down, and this too flows as far as possible through management channels, though in this case there will always be a parallel channel, especially for grievances, through the trade union representatives. GEC compare their management structure to a tree in which the foremen are the roots.

The week by week process of this up and down flow is achieved through the now well established system of briefing groups, the cascade system, in which each manager explains and discusses current and future problems with the group of managers immediately subordinate to him so that at the end each foreman will brief his own working team. This process should be continuous – weekly or sometimes daily – so that information does flow simultaneously up and down. On controversial subjects trade union officials and shop stewards may be informed before those they represent, but not so far in advance that they can preempt the foreman's briefing.

There are two major briefings done as a matter of course each year - usually through briefing groups: one (in June) looks back on the previous year's accounts and performance. The other (in December) looks forward to the next year's budget and plans. These are usually supported with handouts, slides and video tapes, and senior management may be called in as necessary to answer questions.

The cascade system is supplemented by a number of other channels where appropriate. One of these is the 'vertical slice' in which a senior manager briefs a section of his whole department in turn, the others remaining at work. This method has the advantage of pre-empting leaks through the unions which might occur during the rather longer cascade process, but the disadvantage is that those not in the first slice may receive a garbled or polemical version of it before they hear the management's explanation.

Periodically – or if there is some major and controversial matter to transmit – the managing director or a senior manager may brief the whole workforce together but this is rare because the more people hear from their own immediate boss the stronger will be his relationship with them.

The most delicate briefings are those concerned with redundancies and factory closures. Due to contraction of the Telephone Division, the company has in recent years had to close two factories, in Middlesbrough and Treforest. In these cases the handling went wrong and the press got hold of the news before the unions had been informed. Timing is crucial and it is essential to minimize the risk of prior leakage, either to trade union representatives or to the press. In an era of technological development such decisions are inevitable but it is equally inevitable that union representatives will do their utmost to resist them and will regard the arousing of shop-floor opposition before the news is broken as their strongest weapon. For this reason they will demand that closures and redundancies should be discussed with the unions *before* they are announced but the management's aim will always be to transmit the facts first, initially directly and personally to those affected and then (about an hour later) to union representatives. GEC's experience is that news of closures and redundancies is usually less of a surprise than is often supposed since the workforce, being well briefed on the company's position as a whole, can see them coming. If the announcement is well handled, severance pay is now so generous that there are usually enough volunteers to provide the necessary redundancies, though this is more difficult when a factory has to close in an area where there is little or no alternative employment.[4]

GEC, despite being a giant firm, has succeeded in minimising industrial conflict and its consequences in management diversion, frustration of technological development, loss of markets and loss of jobs. While some of this success can be ascribed to the nature of their activity (e.g. easily buffer-stocked products, easy second sourcing of materials, and strong salaried and female elements in their work force) the main explanation has been good communication. This has been made possible by decentralization into autonomous units of manageable size and by highly professional and personal management in which time spent in communication is regarded as saving greater time spent in disputes, and as essential features in creating the kind of shop floor atmosphere which results in high productivity.

17 Japanese Management in the UK

JAPANESE WORKING PRACTICES

The success of Japanese subsidiaries in Britain has surprised many people. They assumed that Japanese managers would treat British workers as if they were Japanese and that the British workers would resent it. Japanese managers have not done this. They have, from the start, accepted the idiosyncracies of British employment and trade union practices and have devised a pragmatic management style to fit in with these practices. The result has been good, efficient and popular management resulting in good *working* practices. These management principles and working practices are, in fact, the same as those to be found in the best and most successful British firms, including some of those firms examined in this book. The conclusion, therefore, is not that we should try to emulate the Japanese in Japan but that we should study the management of British workers under British conditions as Japanese expatriate managers have done here, and use their experience to reinforce the already existing examples of good management in British firms. As one Japanese firm wrote to the author: 'We are, as you will see, a company with largely UK management operating policies which are appropriate to the UK.'[1]

This achievement by Japanese managers is all the more remarkable when one considers the totally different employment practices on which they have been reared. In Japan there are enormous incentives to remaining in the same firm for a whole lifetime and it has been estimated that leaving a good firm often results in a drop in salary of about 40 per cent.[2] A Japanese manual worker receives an annual increment, quite separate from higher pay scales after promotion, which means that a middle aged man doing the same job as a younger man on the machine next to his may receive more than double the younger man's pay. The big firms provide comprehensive welfare benefits, including housing, medical treatment and schools which might arouse accusations of being too paternalistic. All of these incentives lead to a certain submissiveness which is in any case part of Japanese culture. Japanese managers coming to UK, however, are trained to understand the British people and environment, with a view not to changing them but to get the best out of them as they are. They seem to reach conclusions not unlike those reached in successful British firms such as John Lewis or Marks and Spencer, which they too have no doubt studied.

The result is good management, in terms of both organization and motivation, resulting in sales and profits, resulting in good pay and stable employment. Since people like working in a well-managed concern, they respond. There is no mystery about Japanese management; it is just pragmatic and efficient.

They have, however, applied successful working practices which work equally well regardless of the nationality or culture of the workforce. In a valuable and constructive study of Japanese management in Britain, the Policy Studies Institute (PSI) picked out four main elements of these Japanese-style working practices:

> A short-list of four can be discerned: an organized or orderly approach, an emphasis on detail, an over-riding priority attached to quality, and a punctilious sense of discipline. Expressed in this way, the list may not sound impressive, and it is easy to understand why the character of working practices in Japanese firms should have attracted so little attention hitherto. But the workers concerned saw these practices as highly unusual; and in manufacturing at least, as giving their firms a great competitive advantage.[3]

Quality control is built into Japanese production processes, partly through automated checks which eliminate human error, partly through self-checking, cross-checking and meticulous inspection at each stage, but above all by developing a consciousness in every process worker and a sense of shame if his or her product is below standard. Again, there is nothing new in this, and British workers have reacted favourably to it.

JAPANESE-STYLE MANAGEMENT

One of the reasons for their ready acceptance is that Japanese managers, whether expatriate or locally recruited, are trained to get far more directly involved in the shop-floor processes than most British-trained managers are. As a basis, they are better trained technically, whether promoted from the shop floor or as graduate trainees. Senior managers can confidently discuss intricate details of production processes with the operators and can and often do advise them on solving specific craft or operating problems or give a practical demonstration. Everything possible is done to develop a sense of teamwork and a realization that more production at low cost means more money and more jobs.

This teamwork is encouraged in small sometimes symbolic ways. Most Japanese companies provide uniform overalls, with name badges, which everyone wears including the Managing Director. (Again, this is not revolutionary: policemen, soldiers, firemen and many others have done it for years). All start work at the same time. All are paid on the same monthly-salary basis. All eat in the same restaurant.

There is constant discusssion, usually beginning at a *daily* briefing group with the supervisor and his team before production begins. Subjects discussed range from detailed production problems to the company's budget or the effect of a change in the international currency exchange rate (see case-study on Toshiba below).

Japanese firms devote much more time and trouble to personnel selection than most British firms do, and thereafter much more time to training, from managers at every level down to process workers. When they set up a subsidiary in Britain, a large number of expatriates come over to train British managers and workers, while a number of British are flown to Japan for training there. Japanese managers remain until they are satisfied that the British ones are fully ready to take their places. Japanese managers are, in fact, generally very popular with British workers. This may contain an element of prejudice, in that Japanese management starts with the advantage of a high reputation, in contrast to the 'adversary relationship' built up over the years in British industry. The main reasons, however, are undoubtedly that the Japanese managers – and especially the junior managers and engineers – clearly do understand the detail of the work; that they treat their workers as equals; and, above all, that the whole operation is efficiently organized.

The PSI study found that Japanese management in Britain has been far less successful in financial firms (banking, insurance etc.) than in manufacturing. White-collar workers have displayed much more resistance to Japanese-style management than manual workers. This has surprised the Japanese themselves, because the British reputation for conflict in industry had led them to expect much more resistance from manual workers, but they had imagined that white collar workers, and especially graduates, would have the necessary motivation without the need for positive leadership. The main explanation, however, is probably that the more highly educated staffs in business firms had been reared on the British liberal tradition of individualism and resented the guidance given by Japanese managers, wheras the British manual worker, expecting to be treated as an impersonal cog in the wheel, was delighted to be treated as part of the team and to take part in a dialogue. Again, prejudice probably played its part.

British manual workers reactions to Japanese management are well illustrated by quotations from interviews in one of the PSI case studies:

> There's a friendly type of relationship between management and the shop floor. The Japanese would be prepared to work 24 hours per day if required.
>
> They tend to push people a lot, but I consider you can trust them and learn so much from them. The Japanese engineers are very good and will help in any way that they can.
>
> The attitude to work [created] by the Japanese is very good. . . In other firms, people work for the money, but in JEL, people work for the firm. I'd do overtime to finish the job without pay if necessary. If the company benefits, everyone benefits.
>
> They try to involve employees with the running of the company . . . they will tell you what the signs are and how they hope to improve anything that's going wrong.
>
> The chance of promotion is good . . . there is an opportunity every four months to be considered for promotion. This is important, as one strives to improve.
>
> Management and employees are all one instead of being two, instead of being separate.[4]

and there was regret when the Japanese managers began to pull out having trained their British replacements:

> The Japanese have fulfilled all our wishes. They are excellent people to get on with, they work hard, there's none of this British white collar/blue collar thing. A manager will pick up a brush. The Japanese managing director will come along and talk to anyone on first name terms. There's a great equality about it . . . we could have done with the Japanese staying on.[5]

It could be that British workers are still enjoying a honeymoon period with Japanese management and that resistance to it will grow. Sadly, there are likely to be one or two trade union representatives in every firm who will try to reintroduce the adversary relationship for political reasons or because they see no role for themselves without it. In the case study which follows, most of the trade unionists concerned saw the benefits of good management and good labour relations and so did the British managers. Therein lies the best hope for industry in Britain – under British or foreign management.

THE TOSHIBA EXPERIMENT

In September 1980, Rank Toshiba Ltd announced the closure of their factories in Devon and Cornwall with 2500 redundancies, beaten by the recession, high UK inflation and the strength of the pound. Toshiba Consumer Products (UK) Ltd bought out Rank's share and, in May 1981, opened a much smaller factory in Plymouth, more highly automated, to produce colour TV sets. They took on 300 of the redundant workers but, as the old company had been liquidated and its staff made redundant, they were able to select them. In order to ensure their motivation, they showed each applicant a video film of the methods and practices the new company intended to use, so that those who joined did so on that basis. The whole operation was conducted by a British managing director, Geoffrey Deith, in close cooperation with Ray Sanderson, national officer for the engineering memberhip of the electricians union, the EETPU.

Rank Toshiba had had six unions – AUEW, EETPU and FTAT for manual workers and APEX, ASTMS and TASS for clerical, supervisory and technical staff. In the new company, an agreement was signed with the EETPU that they would have the sole collective bargaining rights for all below supervisory grades. There was no closed shop, however, and anyone could join any other union, or no union, if he so wished.

In the agreement signed with the EETPU there were a number of unusual features, all of which have proved very successful during the first two years of the company's operations. The first of these was that the union and workers agreed on fully flexible work arrangements, whereby any worker could be asked to do any job for which he had been trained and the more skills in which he qualified the more highly he was paid – a clerical worker, for example, could learn operating or technical skills and vice versa. There were no production bonuses and just three pay-scales for non-supervisory grades: for those qualified in one skill, three skills or seventeen skills.

The second innovation was the formation of a Company Advisory Board (COAB) (see Table 17.1). This had a function similar in some respects to a German Works Council. Its role was advisory and it was separate from the collective bargaining process. It had 10 members, each representing a sector of management or workforce, elected by secret ballot.

TABLE 17.1 *Toshiba (UK) Company Advisory Board*

Constituency	*Members*	*Number of employees*	*Initial term*
Senior management	1	6	6 months
Management	1	17	2 years
Administration/specialists	1	29	2 years
Supervision	1	9	1 year
Technical/maintenance	1	8	6 months
Panel sub-assembly and adjustment	1	26	18 months
Sub-assembly	1	39	18 months
Panel assembly	1	67	1 year
Machine shop/flat pack etc.	1	43	2 years
Funds/quality control	1	56	1 year
	10	300	

Membership was normally to be for one year but, to avoid simultaneous turnover, the initial members were elected for varying terms as shown.[6]

Union members or non-members were equally eligible to stand. If the senior shop steward was not elected as a constituency member he would be an ex-officio eleventh member. The Managing Director would also be a member and would normally take the chair.

The third feature of particular interest was the procedure for resolving industrial disputes. Though not involved in the bargaining process the COAB could (and usually did) resolve problems before they became disputes, since management and union were both represented. Failing this, normal negotiations between management and shop stewards would follow. If they were unable to find a solution, the Toshiba/EETPU Agreement laid down that the question would be refered to an agreed independent arbitrator, whose decision would be binding on both sides. The exceptional feature (though not unique) is that the arbitrator is barred from compromising. The two sides each put their case and he is bound to accept one or the other. This is known as the pendulum system. The principle is that, if either side puts an unreasonable case the other has only to put a reasonable one to be sure of the verdict. In practice the pendulum system acts as an effective deterrent to disputes ever reaching that stage, since the two sides usually arrive at proposals so close together that they can be bridged by agreement.

An example occurred in the 1982 pay round. Although management and unions moved close, the COAB could not agree. The union held a secret ballot on a compromise proposal put forward at the COAB and this was accepted, probably on the grounds that the company's original offer was so close to the unions' demand that the arbitrator would probably find for the company – which would have deprived them of the compromise gained in the COAB.[7]

In other respects, Toshiba follows the practices already described for other Japanese-managed companies in the UK. Everyone including the Managing Director wears the same uniform, eats in the same restaurant, starts work at the same time, and is paid a monthly salary. The work force is regularly briefed, not only through its COAB representatives, but also direct in briefing groups or what the company calls 'small circles' – a group of, say, eight people taken from an assembly line of 40, to permit more intimate discussion and to avoid halting production.

The company makes a point of not only inviting suggestions but acting on them wherever possible. In 1981, for example, a change in the £/yen exchange rate went badly against the UK subsidiary, subjecting them to the prospect of a loss of £680 000 on the first year's operation. The problem was put to the workforce through COAB and a 'currency counterbalance team' was formed with nominees from each constituency. Everyone came up with the ideas – amongst them one from the soda-bath operators which saved £40 000. These, together with savings by management, eventually covered the entire £680 000. Since then, the current £/yen rate has been prominently displayed throughout the factory, and everyone understands when problems are likely to result from it and is motivated to find ways of helping the company to survive.[8]

Some problems inevitably arise from the conflict of interest for shop stewards who are also members of the COAB. This is not because they wish to see a return to the adversary relationship, because public opinion on the shop floor is overwhelmingly against that. It is, however, difficult for a shop steward to negotiate around a plan to which he has been a party in the COAB. Such problems will occur whenever trade union representatives join participatory bodies, which is why they are reluctant to do so. The original senior shop steward at Toshiba in 1981 was in fact replaced for this reason in 1982 though she remained an ordinary shop steward and a COAB member. The underlying influence at Toshiba, however, is that the work force, having faced redundancy once from Rank-Toshiba, have a stronger incentive to help the new company survive than to strike a hard bargain. The real test of the management will be whether they can maintain this attitude over the years as the honeymoon atmosphere fades.

To quote Ray Sanderson, the EETPU national officer who signed the original agreement in 1981:

> Over the last three years I have been looking very closely at industrial disputes in the manufacturing sector; in the vast majority, our members went back with more or less what they were offered before they came out. And it isn't difficult to understand why. Manufacturing industry is becoming more and more competitive and is likely to remain so. Industrial action is a self-defeating method of resolving conflict and employees have got to identify with the interests of their employers, because that interest is survival.[9]

And on the subject of the applicability of German and Japanese methods to British industry he added:

> You know, I have been to so many international conferences where the British delegates have been very smug about the Germans and the quiescent Japanese, and really we are the mugs, not them, because the German worker is twice as well off as the British worker and the Japanese are 50% better off. With all our militancy and all our sloganising, what we have actually achieved for our members is poor by comparison.[10]

Part IV

What Is To Be Done?

18 Alternative Routes to Industrial Revival

CREATING A WINNING TEAM

The theme of this book has been that Britain must create a winning team within the next 10–15 years by developing the will to win and eliminating those who shoot through their own goal. There is no doubt at all that Britain *could* do this, just as she *could* catch up on modernization of her industry if she were to apply the knowledge and research capability she possesses and apply or attract the necessary investment to do so. That investment however, will only come if both British and foreign investors become confident that Britain has cured the British disease and that the team in which they invest is one which will win.

There is no single magic formula. There are, instead, many proven winning tactics and techniques before our eyes, one or more of which will apply best to each individual industry, company or service. These tactics and techniques will be summarized in turn in this chapter, with reference back to the case studies and examples in the body of the book. Some are already practised in Britain; for others we can best learn from the experience of other countries.

One technique which must, sadly, be dismissed as a failure is our own (1945) pattern for nationalized industries. Some public services must by their nature be provided by the state (such as the civil service, the police, the armed forces and the fire service). For others there is a case for subsidies (such as the railways, in order to reduce the traffic on roads). Generally, however, the nationalized industries have failed to pay their way (though some, like BL and British Steel, are hopefully beginning to do so.) A more serious problem is that they have quite failed to live up to the hope in 1945 that they would be free from strikes because workers would not wish to strike against 'the people'. As discussed in Chapter 11 they have in practice had a worse strike record than private industries. There is, as a result, little public support for further nationalization. There are few people who would want to see, say Boots the Chemist replaced by a 'Peoples Dispensary' or Marks and Spencer by a 'National Clothing Distribution Board'. They believe that they would get poorer service and pay more.

Why have the nationalized industries done so badly? On the management side, the imperative to be efficient and profitable in order to survive has been lacking; and, though there have been some able and dedicated Chairmen, the best chief executives are more often attracted to businesses they can run on straight commercial principles without political intervention. Amongst the work forces, there is an underlying feeling that the Government would not dare let a nationalized industry or service collapse, so that it will be easier to squeeze them for higher wages than a private firm which could go bankrupt. (In fact, many nationalized industries would have gone bankrupt long ago if they had had to rely on attracting private investors or making profit to reinvest.) Clause IV of the Labour Party's Constitution, whereby members pledge themselves to work for the common ownership of the means of production distribution and exchange, has become one if its greatest electoral handicaps. This may be because there is a fundamental difference between state ownership (i.e. nationalization) and collective ownership of an organization by those working in it. State ownership clearly does not work well in Britain, whether the Conservatives or Labour are in power.

COOPERATIVES AND PARTNERSHIPS

The Wilson Government's ill-starred attempt in 1974 to rescue three ailing industries by financing 'cooperatives' with public funds (KME, Meriden and the *Scottish Daily News*) failed because they were really forms of nationalization rather than cooperatives. Certainly in the case of KME most of the work force never felt that they owned it, nor that they needed to make a profit, because the Government would put in more money to bail them out. The KME experiment also proved that combining trade unions and management was a recipe for disaster. The experiment provides a whole range of lessons in how not to do it. Nationalized companies are not workers' cooperatives.

Nor is 'the Co-op' wholesale and retail chain whose 'ownership' by its customers is so diffuse that it differs little in essence from a public limited company.

The great majority of workers cooperatives in Britain are very small, and are not directly financed by Government, though they may qualify for grants or rebates as an incentive to set up in depressed areas, just like any other form of company. The main government sponsoring agency (CDA) acts as a catalyst for raising funds and not as a central bank and it tries to ensure that the cooperatives it supports have a strong enough personal stake (in terms of their money or their jobs) to carry them through and if necessary to make sacrifices to maintain profitability. Given these conditions, and with the growing opportunities for small businesses in the fastest developing industries, there is great scope for more of this type of cooperative and, though it is still a small sector of industry (9000 employees in 1983) it is growing fast.

There is also a need for a British version of the Spanish Mondragon system, which JOL would like to encourage. The Mondragon system, however, depends on a central bank which is committed to the cooperative ideal. In Mondragon this bank has developed one successful medium-sized manufacturing cooperative (Ulgor, with 3500 employees) and 75 smaller ones (57 of them also industrial). Part of the recipe for success has been the insistence that every employee has a substantial personal stake (at least £1000) and forfeits 20 per cent of this and any accrued profits if he leaves the cooperative. Since the bank owns 60 per cent of the capital (generated by its profits) it can and does exercise powerful management discipline. In other words, Mondragon's viability is, and is seen by its employees to be, dependent on commercial success.

In many of the other British cooperatives and partnerships the workers do not have a personal capital stake in the business. though they may well get a substantial bonus to their income when it is profitable. These cooperaties mostly owe their foundation to the generosity of a successful businessman who chose to direct his wealth to an ideal, notably Spedan Lewis, Ernest Bader and Philip Baxendale. In each case, however, these men have devised constitutions which have ensured that the management does retain freedom of action in making commercial decisons, which is the most vital ingredient in any kind of partnership or cooperative.

When Ernest Bader handed over his chemical firm, Scott Bader, to his work force in 1951, his constitution precluded individual workers owning a stake in the capital. Though it remains small (350), Scott Bader has, since then, generated its own capital by its own commercial success. Its success led Bader to found two organizations, ICOM and ICOF, which sponsor and finance small cooperatives on the Scott Bader model, and this, though different from Mondragon, is a model which works. The essential ingredients are: adequate priming capital from a non-government source; a constitution which ensures freedom for commercial decisions by management; and, flowing from this, self-generation of the capital needed for survival and growth.

The success of the John Lewis Partnership was built on similar lines, with these same three essential ingredients and a number of others. Its growth has been even more phenomenal – from £1 million launching capital in 1929 to £320 million in 1982. It has for many years been ploughing back 60 per cent of its profit in reinvestment but the other 40 per cent has been distributed in cash bonuses to the 25 000 partners. These bonuses have been equivalent to the dividends they could have expected from a holding of some £20 000 in shares if they had been shareholders instead of partners. As in Scott Bader, however, they own no capital individually. Above all, the constitution is framed to leave commercial freedom of action to the Chairman subject to the ultimate power of the Partners' elected Councillors to dismiss him.

Many of the Partnership's other strengths – such as their breathtaking freedom of speech in the anonymous correspondence columns of the *Gazette*, would be equally applicable to more conventionally managed companies so they will be discussed further in the context of management communications later in this chapter.

It is still too early to say whether the John Lewis system would work as well in a manufacturing industry, with a strong trade union involvement. Mr Philip Baxendale's decision to launch a similar system in Baxi Heating does, however, show every promise of success.

There will never, however, be more than a trickle of successful businessmen willing to place at their work forces's disposal not only their money but their power. In all the cases quoted their faith has been justified: they have remained Chairmen as long as they wished and have remained personally prosperous but there was no guarantee that they would do so.

Whether this kind of partnership system can be applied more widely in Britain, therefore, depends on whether a formula can be found which would encourage the banks to provide the launching finance and reserves in the expectation of getting a reasonably sure and attractive return on their investment. Perhaps the answer will lie in the formation of a trust (like ICOM and ICOF but on a much smaller scale) to promote and encourage the financing of such partnerships by the joint stock and merchant banks, by devising safeguards to give expectation of an attractive return on capital, thereby convincing the banks that a successful partnership like John Lewis, as a team pulling together, would be one of the most profitable of investments for them.

THE CONVENTIONAL PATTERN DONE WELL

That, however, is a strategem that can only start in a small way and would take many years to develop. Britain must meanwhile put its faith in the wider application of the techniques – above all, of management and communication – which have proved successful in conventionally constituted public limited companies. Several have been examined in this book.

Marks and Spencer are one of the most successful because of their outstanding management–labour relations arising from good communication upwards and downwards. Some of the techniques they use were described in Chapter 14 and are mentioned again in the section on British Management later in this chapter. These good communications are all the more remarkable in a giant corporation (Britain's fifth largest) and are facilitated by their natural division into over 250 autonomously managed stores – a similar advantage to that enjoyed by their fellow retailers, the John Lewis Partnership. The high regard in which the work force hold their company and the material benefits they enjoy are, once again, dependent on the company's continued commercial success and expansion. Efficiency and success are the primary foundations of respect.

Another of Britain's giants, GEC, has achieved similar success and confidence in the industrial field. Once again, decentralization has been a major factor – autonomous units of 200 are regarded as the ideal, with 500 as the maximum. Constituent companies larger than this (like GEC Telecommunications, the subject of study in Chapter 16) are subdivided to achieve that end and it is noticeable that the smaller the subdivision the more harmonious the industrial relations. As in M & S, upward and downward communications in GEC companies are good.

We have much to learn from the success of Japanese management in Britain. The Japanese have, very wisely, not treated British workers as if they were Japanese workers and have applied their management and technical skills to British conditions and to the British mentality. As a result the lessons to be learned from the management of Japanese subsidiaries in Britain are highly applicable to British management of indigenous British firms. Perhaps their greatest achievement is the feeling of teamwork engendered by respect for the professional skill and hard work of the management and by the self-denial of management privilege in such symbolic forms of similar hours of work and similar systems for salary payment, catering and even clothing. Other important ingredients are a respect for the managements's uncompromising insistence on high quality production and frank discussion of both problems and performance of the company at every level. In the main case study in Chapter 17, the excellent relationship established with trade union representatives and the 'pendulum' system of arbitration are particularly worth noting.

WHERE SOME OTHERS DO IT BETTER

The British have throughout their history been reluctant to learn from foreign examples,[1] but they have a great deal to learn about industrial relations from abroad, notably from Germany, Sweden, Switzerland and Australia.

Germany can teach us most and the structures and practices which have been the foundation of her overwhelming victory over Britain in the economic race were described in Chapter 4, so they need only be briefly summarized here. They have only 17 industrial unions compared with 480 in Britain.[2] The DGB (equivalent to the TUC) is not affiliated to any political party. Collective bargaining agreements are negotiated centrally and are legally binding for their duration, normally a year. Strikes in breach of such agreements, or which aim to change the agreements while still in force, are unlawful so that the strikers are liable to heavy damages or dismissal. The same applies to strikes called before conciliation procedures have been exhausted or without a prior secret ballot in which 75 per cent vote in favour of striking – this last requirement at the behest of the unions themselves (all except one), not of the law. As a result, German industry has lost on average only one day for every ten lost in Britain and there is a degree of management – labour accord which has raised German productivity real earnings to roughly double our own. Public service strikes are illegal unless the unions provide proposals for maintaining essential services and full time established public servants are barred from striking at all. All of these regulations and practices are accepted with general approval by trade union members and the public at large.

Germany also provides valuable models for industrial democracy. Their elected Works Councils have a long history, secure legal rights and a strong influence. Central collective bargaining agreements normally fix minimum rates and leave Works Councils free to pursue plant bargaining for increments, which gives them a strong position with the work force. The German two tier system, again with nearly a century of legal status behind it, provides a Supervisory Board with parity representation (though the shareholders' representatives have a casting vote if there is a tied second ballot) in which representatives of junior management, staff and manual workers are directly elected by secret ballot.

The pattern of German trade and industry has more in common with our own than almost any other, so we should draw the appropriate lessons from their success. The theory that German and British national characters are so different that their successful ideas cannot be transplanted has no substance and is used mainly as an excuse by those, both in management and unions, who wish to resist change to their own entrenched and often discredited practices. The biggest difference is one of attitude, not of irremediable differences in character, and the results speak for themselves.

It is clearly more difficult to transplant ideas from Japan, where the entire pattern of employment (e.g. immobility and retirement at 55) and wage structures (e.g. large increments based on age and service for manual as well as managerial workers) are very different. Japan, however, can teach universal lessons regarding communication, participation and, above all, the engendering of a team attitude. These have been encapsulated in *Theory Z* which has swept the US business world in recent years.[3] Fortunately the principles of Theory Z are well demonstrated in British companies under Japanese management (see above) and in good British companies such as the John Lewis Partnership and Marks and Spencer so we do not need to look beyond our own shores to find these models.

Sweden and Switzerland are also worth studying for models of industrial cooperation. In both these countries there are long standing accords between management and trade union organizations, formally enshrined in agreements in 1938 and 1937 respectively, which have amounted in practice to a joint determination to solve disputes peacefully. This attitude is best illustrated by a repeat of an extract from the statement by the Chairman of the Swedish Federation of Labour.

> It is easy to act so that society cannot function. But such a conflict is no longer a conflict against employers but rather against society and ourselves. Society is to a large extent ourselves and our families.[4]

And as an illustration of the Swiss attitude, the Federation of the principal unions voted in July 1983 to accept a reduction both in working hours and in real wages over the next 5 years.[5] Lest it be thought that these attitudes might be tied to a more egalitarian restraint on executive salaries, a survey published in August 1983 recorded that in the 14 leading industrial countries of the world, the average real earnings of higher executives were highest in Switzerland and lowest in Sweden.[6]

Efforts to encourage the adoption of the techniques of the Germans and others for communication and participation are more likely to succeed than attempts to enforce them by passing resolutions in the European Parliament as in the EEC Fifth Directive and the Vredling proposals (see Chapter 5). National legislation such as the Australian Industrial Democracy Bill of 1983, is probably more appropriate for this purpose.

This Bill was passed by the Australian Senate on 26 November 1981. It was introduced as a Private Members' Bill by Senator John Siddons of the Australian Democratic Party (roughly equivalent to the British SDP). It provides for the formation of an Industrial Democracy Board whose task is to encourage companies to qualify for tax incentives by fulfilling certain qualifications:

(a) At least 8 per cent of the share capital to be owned by non-executive employees.
(b) The enterprise to be organized in working groups of not more than 100 members, each group to be autonomous, with wide powers of decision making and appointment of staff, and each to be a profit centre with separate profit and loss accounting.
(c) The enterprise to have a profit-sharing scheme whereby an agreed proportion of pre-tax profits within each group are distributed to employees.
(d) The enterprise to have a consultative council, half of whose members are elected by employees by secret ballot and half appointed by management, the chairmanship to be held by each half alternately.

Enterprises so qualifying would be registered as Industrial Democracy Enterprises and would receive an 8 per cent reduction in Income Tax.[7]

The SDP/Liberal Alliance would have introduced similar legislation in Britain, had they been elected to power in 1983.[8]

BRITISH MANAGEMENT AND COMMUNICATION

Modernization of British industry to make it fit to compete in the boom of the 1990s depends on its generating and attracting much more investment in the 1980s. This in turn depends on securing the confidence of the foreign or home investor that the industry will be profitable, and not subject to disruption by industrial conflict; also that both management and labour will be able and willing to adjust dynamically to changing technology and changing markets. All of these things will depend on good management by imaginative executives.

New plant achieves nothing, however, unless those managing it and those operating it are ready to use it to produce more in less time at lower unit cost. A motivated team with outdated equipment can still beat an apathetic or obstructive one with new equipment.[9]

The essential framework for dynamic management is decentralization and accountability, with clearly recognized profit centres so that each manager knows – and all can see – what he is responsible for managing and how successfully he does it.[10] The John Lewis Partnership *Gazette*, with its weekly publication of the performance of every department, with the name of its manager, provides an excellent example.

A manager can make or lose more than his annual salary in a single day by one right or wrong decision. It is therefore folly to skimp over his selection, his training or his pay. Routes to middle and higher management must tap all sources, from the graduate to the man from the floor who has the ability to apply his experience and to organize and lead. Selection methods must combine science with shrewd human judgement (Sir Michael Edwardes, with his psychologists, used both for his sweeping changes in BL Management in 1977–9). Reliable selection of a management trainee cannot be done in less than two or three days but it is well worth the time. The army's three-day Regular Commissions Board (RCB) provides a proven model (the best of its early selection of 20-year-olds are now generals) which is widely copied in industry. Such three days normally include tests of aptitude (mechanical, mathematical, analytical etc), of written self-expression, of ability to argue constructively in group discussion, of leadership ability in a group and under pressure, and of intelligence; also interviews with psychologists and with experienced managers. It is depressing to see how many firms still rely only on one or perhaps two interviews of one hour or less.[11]

Management training is built on the base of the national system of education and here the conflict of elitism and equal opportunity inevitably conflict. Top managers are, or should be, the most able one out of 100. Put another way, the manager of 800 people will be doing a job which not more than about 7 of those under him are ever likely to reach during their lives. If he (and these seven) are *not* the best fitted of the 800 to do the job there is something wrong with the education and selection systems.

Should the schools have streaming or mixed ability classes? Streaming is clearly the best way of developing the brightest 10 per cent, but will the school selection necessarily have selected the right 10 per cent? And is this worth the price of condemning the mediocre to perpetual mediocrity and the dimmest 10 per cent to perpetual consciousness of – and even a defiant pride in – being the rejects of society? The best compromise is probably to have mixed ability classes up to, say, 14, in which the dimmest will be helped by the brightest, who will themselves benefit from the experience; then for streaming to be introduced in fifth and sixth forms for CSE, O Levels and A Levels.

Most universities (with some notable exceptions) continue to distance their teaching from anything 'vocational' (dirty word!), and especially fail to give regard to the needs of commerce and industry – an astounding tradition in a country whose survival depends upon manufacture, business and trade.[12] It is also important to attract more good students to read science and (more important) engineering. This will not happen unless engineering graduates are seen to have a good chance of reaching top management as they are in almost every European country except Britain, and unless engineers are educated more broadly, especially in economics and business management. This will solve another problem, because engineers will be seen to reach positions in which they receive salaries comparable with those of lawyers and accountants – again as on the Continent and as in USA and Japan.[13] In Britain they seldom do.

The first year of management training is good in most big firms, with a fair taste of the worm's eye view and a broad understanding of the place of each department in the whole. The weakness comes after that, because young managers often go into their first full appointments without sufficient specialist training for the immediate operations they have to manage, so they lack professionalism and fail either to develop confidence or to earn respect from the start of their careers. (The Japanese do not make this mistake; see Chapter 17.) Perhaps the best foundation of all for future top managers in extractive or manufacturing industries is an engineering degree, then three years work getting their hands dirty in the essentials of their firm's productive activity, after which they should move over to, say, marketing or financial management, to broaden their experience and their outlook in preparation for higher management.

Another weakness is in development training for middle and higher managers. Because of real or supposed pressure or competition, firms are reluctant to withdraw good mangers from executive posts for more than two or three weeks whereas, about every 7–10 years, the potential star performers should be taken right away for several months middle and higher management training – as they are in the Police (e.g. the Special Course lasting 12 months for selected high flyers in their late twenties and the Senior Command Course for six months before becoming Assistant Chief Constables at around 40). It is argued that this means financial cover for, say, 10 per cent more high-salaried managers, but it costs far more in the end to have less effective managers in terms of both general results and of the cost of even one or two wrong decisions.

British firms often waste far more than this 10 per cent in extra salaries by giving a manager too narrow a span of control, i.e. with too few people reporting to him from the next level down. Between 5 and 7 is probably ideal, because he is then forced to decentralize, whereas if he has only 3 he not only breathes down their necks, but the total number both of levels and of high salaried managers is far greater than with 5 or 7.[14]

In many of the case studies in this book, 200 has emerged as an ideal for an autonomous profit centre (e.g. GEC, Marks and Spencer). This is the number that the manager of such a group can know personally, not only by name, but also well enough to discuss their jobs, careers, families or other problems knowledgeably and constructively. Where this kind of relationship exists, industrial disputes are almost unknown, and teamwork develops.

One problem, sometimes very hard to solve, is that leadership (charisma, warmth and organizing ability) does not necessarily coincide with high technical ability. In such cases the good leader can master enough technical knowledge to get by, given competent subordinates, whereas the most brilliant technician who lacks the intrinsic qualities may never become a good leader, and it is leaders who produce results.

Many managers find themselves unable to escape from accounts, charts and plans and so do not get out onto the shop floor or the coal face. While it is vital that they do make right decisions, especially financial and commercial ones, they should be able to rely on their subordinates to feed them the material to do so. Those in the direct line from the managing director to the production process itself must above all have a feel for the shop floor and should spend a great deal of time there every day. This again, is something at which the Japanese excel. If productivity is to be high there is no substitute for it.

Various structures for participation and industrial democracy have been discussed in this book but the structure matters less than the attitude. Participation will achieve nothing unless both management and labour wish it to do so and believe that it can. Participation is also sterile without information. There are two hazards: first, the danger of leaks to competitors and secondly the misuse of such information by politically motivated trade unionists whose aim is to 'screw the management' or even to bring the whole operation to a halt. The best guard against both of these is for the workforce to become convinced that either of them will prejudice their success, their earnings and their jobs. Members of a top football team would give short shrift to a traitor who leaked their tactics to a rival before a match. Given this attitude, at least as much should be disclosed to workers as to shareholders.[15]

Of the various method of communication upwards and downwards discussed in the case studies, the 'Briefing Group' concept[16] (under whatever name) is by far the best. There is nothing new or revolutionary about this. It consists simply of each manager from managing director to foreman, discussing his problems regularly with his immediate group of subordinates, both giving and receiving ideas which then go upwards and downwards through the same channel. Nelson did it with his Captains (his 'band of brothers') and they did it with their ships' officers. In the Peoples Republic of China they do it, not only in industry, but in their public administration downwards through city, street and neighbourhood committees. Everything, they say, goes 'up and down' from wherever it starts, to and from top and bottom, until a decision is reached. This is precisely the system operated so successfully by the Japanese. If often takes them longer to reach a decision than in Europe or the USA but when the decision is taken everyone is expected to abide by it loyally – and they usually do because they have discussed it themselves, up and down, before it was taken. When this process has to be telescoped because of urgency, this is explained and, because it is rare, it is accepted. There is no structure of management to compare with this because it does bring everyone in, up and down, from the top manager to the man on the machine. And it has proved effective wherever it has been done.

BRITISH TRADE UNIONS

In 1979, Britain's worst year of strikes for over 50 years, less than 5 hours were lost in strikes for every 1000 hours worked; that is about half of one per cent – a fraction of what was lost from sickness alone,[17] not to mention other causes. Yet this half of one per cent has done more damage to the British economy than all the rest put together for the reasons brought out in this book: consequential layoffs and disruption of other industries (applying especially to public service strikes); perpetuation of outdated methods, outdated plant and overmanning; and diversion of management time and effort from production to conflict resolution (up to 50 per cent of their time in the British motor industry in the 1970s – see page 65). It is these consequential effects which have brought about the dismal state of British productivity and hence the halving of the standard of living which we should have.

One of the reasons for this is that there are very strong incentives for management to give way to pressure for restrictive practices and overmanning; partly just for a peaceful life – to avoid the constant, time-consuming hassle of dispute procedures; and partly for the reason given on page 62 – Sir Alex Jarratt's estimate that strikes cost the employer about 100 times more than they cost the trade unions. Industrial conflict is now (in contrast to the 19th century) heavily weighted in favour of the unions.

In capital intensive industries, the wage bill is only a small fraction – perhaps 10 per cent – of the total cost of production so it usually costs the employer far more *in the short term* to stop production than to give way; so he is tempted, with the imperatives of cash flows, to mortgage his future competitiveness by paying a heavy price for peace. The trade union negotiators know this very well so the half of one per cent of days lost conceals a vastly higher percentage of future productivity lost, future sales lost and, in the end, future jobs lost.

Modern trade unionists know this very well too but, just as surrender may pay the employer in the short term, so militancy pays the convenor and the shop steward in the short term, because he is seen to fight for and win better pay, and thereby strengthens his following. He, too, is tempted to cash in on short term dividends at the long term price of a lower standard of living and future redundancies for those he represents.

Yet, despite the temptation, the majority of trade union officials and representatives do not do this. It may be that the half of one percent in lost days is compounded twenty fold to 10 per cent forfeiture of productivity each year[18] by management giving way 20 times for each time they fight the challenge through to a strike. Even if this were so it would still imply that about 90 per cent of officials', convenors' and stewards' time is spent *solving* problems rather than driving them to the point of going into dispute procedure or stoppage – i.e. in maintaining production and often increasing it to earn their constituents higher bonuses.[19]

While the militant minority do irreparable damage, this constructive majority provide continuous benefit to their factories, services and workforce, by preventing frustrations building up into disputes, by devising ways of increasing productivity (and bonuses) and, above all, by providing leadership to carry their people with them in accepting new plant and new techniques by convincing them that these will pay in the end. This is the constructive role played by 99 per cent of the Swiss, Swedish and German trade union leaders and by a (sadly smaller) majority of our own.[20]

The other cross the British economy has to bear is the political involvement of the trade unions. All our case studies – especially the international ones – bore out the malignancy of this aspect of the disease. In all the most successful countries, Germany, Japan, Sweden, Switzerland and the USA, the unions maintain their freedom of action by being independent of any political party, though they can and do subscribe to party funds, often to more than one party. Ironically, the relatively successful economy in France [21] is partly accounted for by the weakness of the trade unions arising from their excessive political involvement, which results in only about 20 per cent of the labour force being members at all. Workers are thus unorganized and much more exploited than in Britain or in any of the other countries mentioned. This may be reflected in short term economic success but also in France's turbulent history – 10 revolutions in 200 years. In terms of human suffering this may have been a greater price to pay than economic stagnation.

In Britain, where the Labour Party was born by, and is still 90 per cent financed by the trade unions, the party and the unions are mutually destructive. The unions suffer from having a divided aim. The movement tries to use its financial and card-voting dominance (e.g. by sponsoring MPs and in the Labour Party Conference) to influence Labour Party policies and, in its day to day activities, to get the Party elected. In the event, each is a millstone round the other's neck as reflected in the very low public esteem in which both are currently held, even by Trade union members themselves. Even when Labour is in power, the relationship yields little benefit: one of the rare periods in which real earnings declined (1974–75) and the highest rate of lost days from strikes (1979) in 50 years were both when Labour was in power. The unions would be immensely more effective and more popular if, while continuing to subscribe as their members wish, they made themselves totally independent as the German and American unions do. So would the Labour Party. The British electorate clearly dislikes 'union government' as much as it dislikes military government.

Trade union officials and representatives who openly use their industrial power for political ends do the movement and the nation immense damage. This is equivalent to that which, say, bankers would do if they deliberately 'paralysed the nation's economy' in order to bring down a Labour Government. This is another reason for its being vital to separate industrial power from political power. Here the media can and do play a role, by providing good background and investigating and unmasking the abuse of power, whether by financiers, industrialists, government officials or trade unionists.

A more serious problem, however, arises from the role into which trade union officials and representatives have been cast by their history and status. They see their support as being wholly dependent on their being seen to fight management and win concessions. Just as militancy gains short term dividends because of pressure on management to withdraw from the brink and maintain their cash flow, so the militant steward tends to gain more immediate successes than the moderate. It is easy to say that management should ensure that moderation pays at least as well, but the price of giving way, as described above, is too heavy for this to be realistic. The answer must lie in public education, by constantly drawing attention to the long term damage done by driving hard and uneconomic bargains. Here, too, the media can contribute a great deal. The argument was well put by Ray Sanderson (EETPU) in the quotation on page 174.

The problem is that if the shop steward's traditional role, the pursuit of conflict as the source of his power, were removed, he might feel that he had no role at all. If trade union members felt the same, would the unions themselves fade away and their constructive role be lost?

The answer, and the way ahead, was indicated in Chapter 11. The cure for the disease must begin in the public services, initially in the services 'essential to life'. The first stage should be a secret ballot of those working in those services on whether they would prefer their pay to be fixed by an independent review body, with suitable safeguards and rights of appeal, as a *quid pro quo* for a legal undertaking not to strike. If the majority voted in favour (as they almost certainly would) then this should be introduced as a condition of service two years later. Provided that the system worked fairly and well (as it already does for the police and the armed forces) it could then be extended into the other public services and perhaps gradually also into the private sector. Its introduction should still be subject to voluntary choice by secret ballot.

If this system did spread, the role of the shop steward, the convenor and the union official would change; he would become an advocate rather than a negotiator. Perhaps a different type of official would emerge; indeed this is happening already; a substantial number of union officials, and a lesser number of stewards, now have higher education. The steward of the future may have a role more like that of the citizens advice bureau and the full time official more like that of a solicitor, able to use his union funds, no longer to dispense strike pay, but to research the case for fair pay and conditions for his members, and to brief a barrister to argue it.

Perhaps this is the type of union and union official that many workers would prefer, and the type which will prove relevant to the challenge of the 1990s.

19 Britain in the 1990s

FIFTEEN CRITICAL YEARS

The previous chapter looked at the ways in which Britain can develop better approaches to management and labour relations, and especially to cooperation, participation and communication. This chapter examines the social and economic aspects of investment and modernization, with particular reference to the microelectronic revolution which will have transformed almost every kind of human activity between the 1970s and the 1990s.

In the 1970s the significance and scale of this embryonic revolution became apparent, coinciding with the world economic recession, which gave some impetus and incentive to research and development to find ways to achieve greater efficiency. Since 1980 there has been some progress. Industry and the business world have made themselves leaner. Small businesses have developed, with the entrepreneurial outlook and flexibility to seek out new openings and exploit them, led by the immensely encouraging research, development, production and marketing of Sir Clive Sinclair's organization. North Sea Oil has provided both the Government and industry with some economic elbow room – though British governments in the 1970s sadly mortgaged far too much of the current and future oil income by gross over-expenditure and public sector borrowing.

The computer has already revolutionized industry and commerce. It is simply a device for data processing which can say 'yes' or 'no' to millions of questions very fast. As such it is only a development from the punched card, on which there are a specified number of points on which a hole can be punched or not punched. The punched card machine has been used for two centuries to control machines or, for example, to select personnel. The machine is asked, say, to select a man who fulfils every one of 15 qualifications such as (1) under 25, (2) unmarried, (3) in good health, (4) holds a driving licence, (5) is a qualified French interpreter, etc., etc. Only if all 15 have holes (i.e. answer yes) is the card selected. The limitation lies in the number of points at which it is practicable to punch – or not punch – a hole, i.e. the number of questions or *bits* of information the card can hold.

Today, a single silicon chip, a few millimetres square, can contain 250 000 bits of information – or 250 kilobits (250 K-bit). Shortly it will be possible to store one million bits (one M-bit) on one chip.[1] When a number of these are linked, the number of decisions becomes astronomical – though each one individually is still only a straight yes or no. The great strength of the computer is that the answers to millions of questions pass through all the circuits involved within a fraction of a second. Another principle is that of the grid, like the squares on a map. There is only one point on a map which will coincide with grid coordinates along and across. Similarly, there is only one point on an electronic 'grid' which will pass impulses along two intersecting lines – one 'yes' amongst hundreds of thousands of 'nos'. A computer can thus take account of literally millions of factors in turn and produce its answer instantaneously – but only if it has been programmed to answer the correct questions. It is a faithful but unthinking slave and, if asked the wrong questions, can give hilariously stupid answers.

It does not, therefore, remotely match the human brain, though the brain does also operate by electrical impulses, containing about 1000 billion electrical connections. To pack in this number of connections when computers were operated by valves would have required a machine the size of Greater London. Built with 1960s transistors it would have come down to the size of the Albert Hall. With the microprocessor (a miniature computer with all its circuits on a single chip) and current integrated circuits, including power supplies, it would go into a small room. The next generation of integrated circuits, already in the science laboratories, may make it possible to concentrate these 1000 billion electrical connections within a machine comparable in size with the human brain.[2] It will still not be comparable in capability (e.g. it will not be able to make emotional or moral judgements) but it will be able to take an immensely complex series of interlocking decisions instantaneously.

It is these factors of size and speed which are producing the revolution. It has made it feasible to produce table sized word processors, and robots smaller than a man to do a man's routine manual work. This routine work is now extending to such activities as inspection. There is, for example, a robot which carries a penetrating electronic eye along a weld; if there is a fault it will spot it, put down the eye, pick up a welding torch and put it right. The tactile robot is also under development, to spot anything which *feels* wrong to its touch.

Microelectronics are not, or course, the only areas of development. The Japanese Ministry of Internation Trade and Industry (MITI) recently spotlighted fibre optics, biotechnology and new materials, along with microelectronics, as the main growth areas during the remainder of the century.[3]

These developments make certain changes inevitable in the established industrial nations. Employment in their heavy industries will continue to decline, the more so as those same industries burgeon in the rapidly developing 'middle income' nations in Asia and Latin America.[4] Research by ASTMS has predicted that technological developments over the 15 year period, if we stuck to the present working week, would enable the 475 000 jobs in heavy engineering in Britain in 1980 to be done by 200 000 fewer people by 1995; and that similar reductions would be possible in vehicle manufacture (371 000), textiles (354 000) and 500 000 in banking and insurance.[5] This does *not*, however, mean that another 1 ½ million people need be unemployed but that the same number of people will be able to work shorter hours to produce more, as will be further discussed later in the chapter.

The design process has also been revolutionized by computer aided design (CAD) which enables the designer to take instantaneous note of every factor which could influence his design, which would in the past have meant hours or days of hunting through files and other data. Now the computer will sound the alarm if any of the factors in its memory indicates 'no'.[6] In 1980 the Institution of Production Engineering in London estimated that by 1990, 25 per cent of British companies will use CAD for component design and 5 per cent for tool design as well; and that, by 1995, 25 per cent of machine tools will have sensory devices for inspection.[7]

The opportunity to reduce working hours will be no less dramatic – perhaps more so – in the white collar industries and those which manufacture equipment in the information field. Already, the telecommunications firms have in recent years cut jobs by one third while doubling their output.[8] Microprocessors have enabled at least one local authority to halve its typing pool while increasing its output by 300 per cent and the typists found the work more satisfying.[9] Reduction of working hours, with more people sharing the work, would make it more satisfying still.

The word processor itself, however, may already be approaching its peak. Increasingly, machines can talk to machines so even fewer operators will be required. Supermarket terminals, for example, in which microprocessors can already check the validity of a customer's credit card, may soon use it to debit the customer's bank account and credit that of the supermarket on the spot.[10] If this extended throughout the retail trade and other services (such as airports and railways stations and to the meters of gas, water and electrical supplies) bills and receipts may largely disappear – and with them those whose job it is to process them.

People whose job is to communicate and to obtain, process and dispense information will in future be able to work with a computer terminal at home, and television conferences will become commonplace. The implications for office accommodation and transport are evident and, in the long run, beneficial.

There may well be other revolutionary ideas in process of discovery in the laboratory or of development in the engineering workshop of which we are as yet unaware because researchers and developers do not like to reveal their ideas until they have a commandng lead. The pace of development is now such that a single invention can become a significant factor in industry within 10 years and dominate it in 20 years. The transistor was discovered in 1948, significant in the 1950s and dominant in the 1960s and 1970s. The microchip (invented in 1969) and the microprocessor (1971) are already significant in the 1980s and will be dominant by the 1990s. We cannot therefore predict what other breakthroughs will come. Will the engineering problems of generating power by nuclear fusion be solved in this decade or the next? If so, there need never be another energy crisis. Are there ideas in other fields (e.g. in biotechnology) round the corner whose nature and significance we cannot even guess? All we can say is that we must be flexible, both in outlook and in the structure of society and its activities; and that every invention will make certain jobs obsolete, whether or not it directly creates others. In both spheres, managements tend to be slow to adapt, and trade unions often greet new developments with suspicion and resistance.

Can we adapt? We have already noted that Britain has adapted from 92 per cent to 2 per cent of the population working in agriculture – and grows more food. She adapted, despite the Luddites, to the power looms and steam engines of the 1800s and to the railway transport revolution in the mid 19th century. The adaptation was a continuing and often predictable process – until some new joker comes out of the pack.[11]

In broader terms, we have progressed from an agricultural to an industrial economy, then to a post industrial service economy and now (according to Barry Jones) we are entering a post service economy. He predicts that, while there will be a sharp reduction in the number of people at work, especially in manufacturing, market-based service employment, and personal services which can be computer controlled (e.g. telecommunications and printing), new types of employment will be generated which are complementary to rather than dependent on new technology. He sees mounting tension between the new "them and us" – the "information-rich", who are employed and affluent and the "information-poor" who are bored and frustrated despite having an adequate income.[12] If he is right, how best can we adapt?

PREPARING FOR THE 1990s

Most of the many people who have thought and written about these problems fall into one of two broad categories – pessimists and optimists. The pessimists see the microelectronics revolution in terms of mass unemployment, assuming that present work patterns will continue. The optimists see nothing alarming about labour saving technology which is in principle nothing new – though the degree and rate of change is greater; they believe that microelectronics can create more wealth with less effort and drudgery and more time and resources available for a more rewarding life-style.

Philip Sadler, Principal of Ashridge Management College, strikes a balance between the two; he believes that we can adjust but that we may fail to do so. His first condition for success is a reasonably stable economic growth in Europe. The EEC has recognized this problem:

> The risk for the Commmunity is that, if we do not innovate as quickly as our industrial competitors, we will lose jobs more rapidly than them, be forced to increase our purchases from them and, by so doing, increase their employment and income at the expense of our own.[13]

The same comment would apply precisely to Britain in relation to her competitors within the EEC.

Given reasonably stable economic growth in Europe, Sadler considers that employment prospects will depend on four main factors: the restructuring of our economies towards more advanced technology-based industries; developing employment opportunities in the service industries; matching education and training to match future job opportunities; and restructuring our employment pattern including school leaving and retirement ages.[14]

Taking the first and last of these: to compete we must quite evidently produce more at lower unit cost, and share the work so that each person works fewer hours but produces more, by harnessing the new technology to better design and productivity. The Japanese faced and solved the same problems before the British did. At the height of the BL strike in November 1981 Mr Ichiro Shioji, leader of the Confederation of Japan Automobile Workers, remarked that Nissan had been on the verge of closing down as a result of a disastrous 4-month strike in 1954.

> The strike was led by a Communist on the basis of a class struggle. Nissan was on the brink of bankruptcy and the management was planning to lay off 2000 of 8000 employees. Instead we all volunteered to take a 25 per cent wage cut and increase productivity if everyone retained their job. The company has not lost one hour of production from labour disputes since then and today it employs 60,000 people. . .
>
> The leaders of trade unions in Westerm automobile factories are partly responsible for unemployment because they refused to accept new technology and robots as quickly as the Japanese. Japanese workers love their robots. They take the drudgery out of work. . .
>
> When we began to replace men with robots in the 1970s production increased. The increase in production was channelled off in the form of exports. This increased production and expanded business. In this way no-one lost their job. Some were retrained in other jobs.[15]

To develop this attitude, it will be necessary to build on the principle that everyone, from management to shop floor, receives a reward commensurate with his contribution. In this respect the Chinese have much in common with the Japanese. It was Mao Zedong who revised Karl Marx's famous dictum to read: 'From each according to his ability; to each according to his *work*.'[16]

This will require piecework or bonuses to encourage operators and maintenance men to get the best out of their equipment. Skill must be encouraged by rewarding flexibility (see the Toshiba example in Chapter 17, where workers are encouraged to acquire 17 skills and rewarded accordingly). Managers also should share in bonuses earned by the profit centres for which they are responsible (see the Australian proposal on page 184). And, while increased wealth should enable differentials to continue to narrow as they have been throughout the 20th century, successful leadership and, in particular, the seizure of opportunities, should be rewarded. Good pay for a successful manager is seldom resented by those he manages and the present British manager/shop floor pay ratio is one of the lowest in the world.

Every encouragement should be given by government, during these 15 transitional years, to the development of small businesses and cooperatives. Mrs Thatcher's Government is doing this and she has a Minister assigned specifically to the task. The biggest problem is to attract venture capital from Britain's generally cautious investors. Tax incentives may be the best way of doing this, rather than direct government investment which usually has a deadening effect. The idea of University Parks, to encourage imaginative entrepreneurs to set up small new technology companies in communities adjacent to the seats of academic research, has been very successful in the USA and in, for example, Cambridge and Bristol. There was a net gain of nearly 18 000 firms in 1981–82[17] and this is an encouraging sign.

This leads to Sadler's second point: education and training. The thrust must be to educate and retrain the national work force for continual transfer from manual jobs to knowledge-based jobs and to the growing information sector so that, within a generation, most manual work can be done by robots programmed and guided by people with the knowledge and judgement to do so. To achieve this, there should be national finance and encouragement for all (other than the dim or unwilling who may prefer to opt out) to stay at school and university for longer. This would have by-products in that they would enter the labour market later and education itself in labour intensive.[18]

This encouragement and finance, however, must be applied to people willing to be educated to fit the jobs required, at the same time teaching them to be flexible and adaptable, to search and to challenge, and to seek and seize opportunities. These qualities must be seen to earn rewards.

Adult education should be more readily available, with an expanded Open University and technical college equivalents which people can drop into or out of throughout their lives using the greater leisure time available,[19] but no one must be allowed to forget that these services can only be financed by more production at a competitive price for less cost. This attitude must be developed throughout the education of the child and through adult life.

The need for continuous education of managers was discussed in the previous chapter, to keep them up to date with changing technology and to develop their management skills. Courses measured in months rather than weeks, every 10 years or so, should become a normal practice as working hours decline and the production of real wealth increases.

Women should be encouraged to bring up their own children by regarding this as work to be paid for like any other. This could probably be achieved best by a combination of negative tax and much larger child allowances, both paid directly to mothers. Again, the smaller demand for productive work and the greater wealth of the community should make this feasible.

Barry Jones suggests a guaranteed minimum income for all, to which they would be entitled whether they worked or not, and that work itself should be voluntary.[20] This is not a good idea as it stands because, the human character being as it is, it would cause strains and resentments in society. Nor is it healthy for the state to be an open-ended paymaster for those unwilling to contribute to the community. Unemployment benefit should be conditional on acceptance of work in the person's own field if it is offered (as now) or, should such work not be available after a stated period, on acceptance of pool work for the community or retraining for a different field of employment.

Jones does, however, make a valuable attempt to rationalize the elusive 'Industry X' which will absorb the surplus human activity, and he envisages a 'quinary sector' (the first four sectors being extractive, manufacturing/construction, tangible economic services and information processing). His quinary sector is essentially domestic activity – unpaid work or home-based work or services where remuneration is incidental. This will include voluntary work for the community, DIY work in home or garden, and similar work for neighbours possibly in exchange for services provided by them in return.[21]

WHAT BRITAIN COULD BE LIKE

In the introduction to this book an indication was given of what could become of Britain if she were to fail to adjust to the challenge of the new technology and to compete successfully with those countries which do adjust to it: the choice between bankruptcy and one or another form of near slavery, i.e. as cheap undercapitalized labour for foreign masters or as pawns in a right or left wing dictatorial regime resulting from an internal revolution.

This final section assumes that Britain does face and overcome this threat to the survival of her freedom, as she has done in the past, and looks at the Britain which could emerge if she does share in the boom of the 1990s.

One writer predicts that:

> It is highly probable that by early in the next century it will require no more than 10 per cent of the labour force to provide us with all our material needs – that is, all the food we eat, all the clothing we wear, all the textiles and furnishings in our homes, the houses themselves, the appliances, the automobiles and so on. In the near future, the vast majority of labour will be engaged as information operatives.[22]

Forecasting a similar percentage in manufacturing industries, Sir Clive Sinclair has estimated that we must create 7 million new jobs by the 1990s, 4 million of these being entirely new and 3 million to absorb those now unemployed. He considers this can be done, and that the jobs themselves will be more interesting.[23]

The new jobs would be created by working shorter hours. Accepting these two forecasts as a basis for planning, this would mean that the average working week in Britain would be 30 hours. If this proved inadequate to keep unemployment reasonably low (it can never be nil), then the working week might have to come down lower still. This, generally, is the solution favoured by the trade unions but it will only work if their members accept that they must produce, not just as much, but *a lot more* in 30 hours than they now produce in 40, in order to provide the surplus wealth for the increased leisure and other services. Examples like those quoted earlier in this chapter (of 30 per cent fewer telecommunications workers producing 100 per cent more or 50 per cent fewer typists producing 300 per cent more) suggest that this is wholly feasible.

Assuming that only 10 per cent of total work is in extractive and productive industries, the machinery will need to be kept working most of the time, ideally round the clock. Thus, allowing time for maintenance shifts in the week's 168 hours (7 x 24), five shifts each working 30 hours (150 hours) might be needed to get the most out of it.

This round the clock (or near round the clock) working of the new equipment will be necessary for three reasons. First, the equipment will be expensive and must pay its way; (the present generation of robots can pay for themselves in manpower saved in 8 years only if they work 16 hours per day.[24] Secondly, unit production costs must be kept low to remain competitive. Thirdly, in an era of development, the equipment may become obsolete or uneconomic before it is worn out.

So there will be fewer factories in Britain, working longer with more shifts. Little or none of the work should be routine and we should share the Japanese aim of all jobs being knowledge-based by the end of the century.[25]

Many more people will work in service industries, including leisure industries, than in manufacturing. This will be more labour-intensive than productive work but again it should be possible for most to average a 30 hour week. Much of the information-based work will be done at home using computer terminals with visual display units (VDUs).

Provided that we do create the surplus wealth, it will be possible to finance that proposed improvements in education (i.e. longer for those who wish it and opportunities for continued education throughout adult life). Retraining courses to adapt to job changes will cease to be unusual. Quite a lot of the education, especially adult education, will be done at home with VDUs, supplemented by 'summer schools' as with the present Open University courses.

More time at home, for work and leisure, will be a boon. As Shirley Williams has written: 'Microelectronics offer the opportunity of reuniting the family and making commuting an obsolete and unnecessary activity.'[26] Less pressure to be within commuting range of big cities will mean that more people can live and work in small towns and villages. Thus local communities, too, will become more integrated and active.

This will provide the basis for developing Barry Jones' 'Quinary Sector'. With more spare time, more people will wish to grow their own food, paint their own houses and exchange such services with their neighbours. Much of this will be on a barter basis, unrecorded for the GDP or for tax. Those who provide what other people need will themselves have a call on the skills these other people can provide – a thoroughly healthy situation in a small community. 'Moonlighting' will fill many gaps in community services and enrich life for everyone.

There will almost certainly be continued pressure to lower the retirement age – perhaps to 55. Here we should turn to Japan and China for guidance because both have traditionally adjusted their family life to this.[27] It is normal for grandparents and great grandparents to stay in the family so it has become customary for them to provide useful services about the house, looking after children etc. In China, old people are only taken into institutions if there is no member of their family available to accommodate them or if there are essential medical reasons. Early retirement may, however, be the most difficult of all the adjustments to make in Britain. Perhaps the new pattern of people working at home and living in smaller more stable communities may revive the extended family with beneficial results for all.

It should certainly be possible for more mothers to bring up their own children, for which they should, as stated earlier, be paid (through negative income tax and child allowances) instead of taking up one of the 30-hour-per-week job vacancies. Communities should become geared, however, to give mothers more leisure and variety in their lives than many have now. The 'Quinary Sector' – and husbands with more time to spare – should be able to provide for this.

This should all add up to a much more rewarding life, with less wear and tear from long hours of working and commuting, more education both for leisure and for work, more family and community activity and a more varied and enriched quality of life.

All of these things, however, are dependent on the actual wealth created in the productive industries and by those international services which provide invisible exports. It is nonsense to imagine that the vastly increased leisure activities can be provided without a massive increase in productivity to remain competitive and to create the surplus wealth. We have already noted that the proportion of the nation's man-hours devoted to wealth production (i.e. extractive and manufacturing industries) will fall from 29 per cent to 10 per cent, so one-third as many people, each working only three quarters as long, will probably need to produce more than twice as much *and to sell it*, on the home and export markets. We shall achieve this only if our production teams are united with a clear aim. Participation and good communication will help to ensure that everyone in the team knows the unit cost which has to be reached if the product is to sell (as in the Toshiba example, Chapter 17); knows how much can therefore be devoted to wages; and knows that it is futile to hope that this wage rate can be raised except by producing more at lower cost and selling it. These truths will apply as much in 1995 as in 1985.

The last word can appropriately go to the Prime Minister who presided over Britain's previous boom, Harold Macmillan. In a long interview on BBC television in 1979, to celebrate his 85th birthday, he said

> The problem isn't going to be work, the problem is going to be leisure. We ought to be thinking about making the machines work for us instead of being their slaves. Not long hours of overtime but many short shifts to make the machine work, keep the machine 24 hours at work, to make it work for us.[28]

Notes and References

CHAPTER 1: BRITAIN'S PLACE IN THE WORLD

1. *Sun* 13 September 1982.
2. Source: *UN Statistical Yearbook*, 1976 cited by John Hey, *Britain in Context* (Oxford: Blackwell, 1979) p.28.
3. Based on figures from Department of Employment *Gazette* extracted by John Hey, op.cit., p.106.
4. This incident was witnessed by the author, in Nienburg, Lower Saxony in 1951.
5. Comparisons are based on figures from official sources gathered and collated by John Hey, op. cit.
6. Sir Clive Sinclair, in BBC TV programme *Futures*, 7 October 1982.
7. This view was borne out in the energy crisis in 1973–74 when industry was restricted to a 3-day week. Production fell by only 25% and it quickly picked up again, many firms getting back to full 5 days production in 3 days before the crisis was over. See page 105.
8. John Garnett, *The Work Challenge* (London: The Industrial Society, 1978) p.1.

CHAPTER 2: THE BRITISH INDUSTRIAL RELATIONS STRUCTURE

1. K. Coates and T. Topham, *Trade Unions in Britain* (London: Spokesman, 1980) pp.32–35.
2. H. A. Turner, *Trade Union Growth, Structure and Policy* (London: Allen & Unwin, 1962).
3. A. Shadwell, *Industrial Efficiency* (London, 1906) p.307.
4. For example, in one of GEC's constituent companies the position is:

Union	*Shop Stewards/Representatives*	*Members*	*Ratio*
TASS	30	508	17 to 1
ASTMS	67	1,000	15 to 1
ACTSS	19 }	950	35 to 1
APEX	8 }		
TGWU	98 }	4,800	22 to 1
AUEW	120 }		
EETPU	3	60	20 to 1
FTATU	1	5	5 to 1
Total	346	7,323	21 to 1

CHAPTER 3: GOVERNMENT INVOLVEMENT IN INDUSTRIAL RELATIONS

1. Lord Donovan, *Royal Commission on Trade Unions and Employers Associations 1965–68, Report*, Cmnd 3623 (London: HMSO, 1968).
2. Draft White Paper, *In Place of Strife*, Cmnd 3888 (London: HMSO, 1969).
3. For a fuller analysis of background, operation and failure of the Industrial Relations Act see Eric Wigham, *Strikes and the Government, 1893–1974*, (London: Macmillan, 1976); A. W. J. Thomson and S. R. Engleman, *The Industrial Relations Act* (London: Martin Robertson, 1975) and Richard Clutterbuck, *Britain in Agony* (London: Penguin, 1980).
4. *The Times*, 23 October 1982.
5. *Guardian*, 13 and 14 June 1983 and Labour Research Department *Fact Service*, 18 June 1983.
6. The June 1983 Election illustrated the anomalies of the British 'first past the post' electoral system when there are three parties with substantial support instead of the two party structure under which it evolved. The Liberal–SDP Alliance, with 26% of the vote got only 3½% of the seats. Labour with 28% of the vote got 32% of the seats and the Conservatives with 42% of the vote got 61% of the seats. The President of the NUM, Arthur Scargill claimed that, with only 42% of the popular vote, Mrs Thatcher had no mandate for her proposed reforms which should therefore not be accepted by the unions. *The Economist*, 16 July 1983 commented that, on that basis, the unions should be denied all the privileges of the Employment Protection Act of 1975, since that was passed by a Labour Government elected by only 38% of the voters.
7. Department of Employment White Paper, 12 July 1983
8. For a critique of the Government's Proposals see *Labour Research*, August 1983.
9. *The Economist*, 16 July 1983.
10. Ibid. For an outline of The Taft Hartley Act see pp.42–4.
11. *The Economist*, 10 and 17 December 1983 and 21 January 1984.

CHAPTER 4: GERMANY

1. Theo Pirker, *Die Blinde Macht* (Munich, 1960) p.17 cited in Kurt Sontheimer, *The Government and Politics of West Germany* (London: Hutchinson, 1972) p.103.
2. Robert Edwards, MP, in an interview with the author in Strasbourg, November 1980.
3. In addition to the Works Council there are, in some plants where the stronger unions are represented, 'Vertrauensleute' (Union trustmen) are appointed to represent the unions at plant level. These trustmen, unlike the Works Council members, have no legal status and their primary task is to advise their members on wage levels, health and safety etc. Some unions have organized 'shop committees' of these trustmen and their presence is sometimes specifically recognized in collective agreements. See Commission on Industrial relations, *Workers Participation and Collective Bargaining in Europe* (London: HMSO, 1974).

4. Collective bargaining sets *minimum* wage levels and collective agreements can and do provide that higher rates may be negotiated at plant level. This has led to some wage drift and also strengthens the Works Councils vis-á-vis the unions in the eyes of employees. See ibid.
5. Stoppages per thousand employees in 1969–78 were 472 per year in Britain and 53 in Germany. Figures from the ILO Year Book cited in Stephen Creigh, Nigel Donaldson and Eric Hawthorn, 'Stoppage Activity in OECD Countries', *Employment Gazette*, November 1980.

CHAPTER 5: SWEDEN, FRANCE, USA AND THE EEC

1. Swedish Trade Union Confederation, *Landsorganisationen i Sverige* (Stockholm, Tiden, 1964) p.18 cited in M. Donald Hancock, *Sweden: the Politics of Postindustrial Change* (London: Dryden Press, 1972) p.151.
2. LO i Sverige, *Kongresspratakoll* (1971) p.745 cited in K. H. Cerny (ed.) *Scandinavia at the Polls* (Washington, DC: American Enterprise Institute for Public Policy Records, 1977).
3. *New Left Review*, September 1975.
4. Sources for compilation of Table 5.1
   (a) Department of Employment, 'International Compilation of Stoppages' *Employment Gazette*.
   (b) *Statistik arsbok for Sverige* (Stockholm, Statistiska Centralbyr̂on 1970) p.236 cited in M. Donald Hancock, op cit.
   (c) John Hey, *Britain in Context* (London: Blackwell, 1979)
   (d) Stephen Creigh, Nigel Donaldson and Eric Hawthorn, '*Stoppage Activity in OECD Countries*' in *Employment Gazette*, November 1980.
5. Cited in Val Lorwin *The French Labour Movement* (Cambridge, Mass: Harvard University Press, 1954) p.30.
6. See White Paper *Trade Union Immunities*, Cmnd 8128 (London: HMSO, 1981) pp.101–3.
7. In Britain, for example, no one has been killed in clashes between strikers and non-strikers or police since 1911 and only five in the period 1893–1911, the most violent since 1842. By contrast, during the three years 1902–4 about 200 people were killed and 2000 injured in violence arising from strikes and lockouts in the USA. See Richard Clutterbuck, *Protest and The Urban Guerilla* (New York: Äbelard Schuman, 1974) p.5.
8. European Movement, *Facts*, November/December 1982.
9. European Communities Commission *Background Report* 15 April 1982.
10. European Movement, op. cit.

CHAPTER 6: THE MEDIA AND INDUSTRIAL CONFLICT

1. TUC, *A Cause for Concern* (London: TUC, 1979) and *How to Handle the Media* (London: TUC, 1979). Other publications criticising this alleged bias include Peter Beharrel and Greg Philo (eds) *Trade Unions and the Media* (London: Macmillan, 1977) and the Glasgow University Media Group *Bad News* (London: Routledge & Kegan Paul, 1976) and *More Bad*

News (ditto 1980); also Denis MacShane, *Using The Media* (London: Pluto Press, 1979).

2. For a fuller account of this strike and the media coverage of it see Richard Clutterbuck, *The Media and Political Violence* (London: Macmillan, 1983).
3. The BBC did some research after the 'Winter of Discontent', later published as *Coverage of the Industrial Situation in January and February 1979* (London: BBC, 1980). This revealed that, despite allegations of bias, more broadcast time was given to interviews with trade unionists than with employers, including government officials (as the strikes were mainly in the public services). On the other hand, some instances were reported in which the attitude of interviewers seemed to be hostile to the trade unionists which might, as with the print media, reflect what the interviewers thought was in keeping with the public mood.

CHAPTER 7: COSTING INDUSTRIAL DISPUTES

1. John Galsworthy's play *Strife* (1912) captures the flavour of the hardship and the heroism superbly but bears little relation to modern strikes.
2. *Strike Trends in the United Kingdom*, CBI Staff Working Paper (unpublished, 1980).
3. See, for example *Financing Strikers* (London: Macmillan, 1977), *The Effects of Strikes on Individual Strikers* (London, Social Science Research Council Report, May 1978) and 'Doling it out to Strikers' in *Personnel Management*, November 1979.
4. See, for example, CBI, op.cit., The British Institute of Management, *Funds available to Employees on Strike*, 1978 and the Engineering Employers Federation, *Financial Support of Strikers*.
5. CBI, op. cit.
6. 'Doling it out to the Strikers' in *Personnel Management*, November 1979.
7. Gennard's article went into considerably more detail and grouped the figures into 'pre-strike', 'in-strike' and 'post-strike' sources of income. In this table the figures have been regrouped to indicate the basic sources themselves, i.e. the state, the unions, others (gifts, etc.), earnings during the strike and at the irrecoverable expense of the strikers.
8. For an account of the Dunlop strike see Chapter 8.
9. BIM, *Funds Available to Employees on Strike* (1978).
10. The employer cannot legally withhold pay in arrears although, if the strike includes clerical workers, it may be physically impossible to pay it out. This applied to the Dunlop strike (see also Chapter 8) and often occurs in others.
11. S. W. Creigh, 'The Economic Cost of Strikes' in *The Industrial Relations Journal*, Spring 1978 also H. A. Turner, 'Is Britain Really Strike-Prone?', Occasional Paper no. 20 (1969), Department of Applied Economics, Cambridge, 1969.
12. Economic League Leaflet no. 5 of 1978.
13. *Daily Express*, 24 November 1978.
14. *Daily Telegraph*, 1 February 1979.
15. Industrial Research and Information Service (IRIS) July 1980.

16. Economic League Leaflet no. 11 of 1980.
17. *Financial Guardian*, Tuesday, 1 December 1981.
18. Creigh, op. cit., p.19.
19. G. Cyriax and R. Oakeshott, *The Bargainers*, 1960, p.114 cited in Creigh, op. cit., p.19.
20. See Chapter 11.
21. 'The Real Cost of British Strikes' in *Now!* Magazine, 16–20 November 1979, p.76,
22. See page 6 and John Hey, *Britain in Context* (Oxford: Blackwell 1979) pp. 12–13.
23. Figure ascribed to the Engineering Employers Federation in 'The Real Cost of British Strikes' in *Now!* Magazine, 16–22 November 1979, p.74.
24. Economic League Leaflet no. 11, 1980.
25. *Strike Trends in the United Kingdom*, 1981.
26. Creigh, op. cit., p.23.
27. Lord Donovan (Chairman), *Royal Commission on Trade Unions and Employers Associations*, Report, Cmnd 3623 (London: HMSO, 1968) p.112.
28. Central Policy Review Staff (CPRS), *The Future of the British Car Industry* (London: HMSO 1975) p.99.
29. CBI, op.cit., p.73.
30. Ibid.

CHAPTER 8: DUNLOP: THE KNOCK-ON EFFECTS OF A STRIKE

1. For an analysis of the bacground of Britain's economy, inflation and unemployment in 1975 see Clutterbuck, *Britain in Agony* (London Penguin, 1980) ch. 14.
2. *Fact Service*, vol. 38, Issue 3, 17 Jan 1976.
3. Ibid., vol. 44, Issue 3, 16 Jan 1982.
4. *New Statesman*, 30 Oct. 1981.
5. *The Sunday Times*, 21 May 1982.
6. The actual figures were £2485 in 1975 and £4483 in 1981 which was equivalent to approximately £1900 at 1975 prices – a fall of 25%.

CHAPTER 9: INDUSTRIAL CONFLICT IN A MULTINATIONAL SUBSIDIARY

1. This case study is based on a series of visits to management, unions and the shop floor in 'APC(UK)' between 1981 and 1983; also on published comments on the events described in newspapers, broadcasts and books. As the company is being described under a pseudonym, however, it is not possible to give the normal references without revealing its identity. It must be stressed that the great majority of the material (often conflicting) came from discussion with individual managers, convenors, shop stewards and workers in 'APC(UK)' who devoted many hours to assisting the case study.

CHAPTER 10: THE MOTOR INDUSTRY

1. Central Policy Review Staff (CPRS), *The Future of the British Car Industry* (London: HMSO, 1975), hereafter *CPRS*.
2. The three sources for this table are *CPRS*, p.108; the Labour Research Department, *Fact Service*, 16 January 1982, p.10, and the Society of Motor Manufacturers and Traders (SMMT) in a letter to the author's research assistant, 10 August 1982.
3. SMMT, op. cit.
4. 'Motors-end of the road for jobs' in *Labour Research*, April 1982.
5. Figures based on total employment, manual and staff, including those in component and parts industries and adjusted to reflect differences in product mix. Sources, (for 1973) *CPRS* and (for 1955 and 1965) C. Patten and A. Silbertson, *International Comparisons of Labour Productivity in the Automobile Industry, 1950–65* (London, 1967).
6. *CPRS*, p.81.
7. Ibid., p.xii.
8. CBI Staff Paper 'Strike Trends in the UK' (unpublished, 1981) quoting figures from the Department of Employment *Gazette* and O. T. B. Smith *Strikes in Britain* (London: HMSO, 1978).
9. Averaged from figures in *CPRS*.
10. *CPRS*, p.99
11. Figures given by Joseph F. Engelberger, President of Unimation Inc, in a lecture to the Young Presidents' Organization Pacific Area Conference at Jasper Alberta in September 1982. Repetitive work in which the pace is dictated by the machine had a 'boredom rating' of 207, the average being the work of a maintenance engineer (100). The least boring jobs were those of the physician (48) and the university professor (49).
12. Anthony Sampson, in *The Changing Anatomy of Britain* (London: Hodder & Stoughton, 1983) p.354. He says that Stokes was persuaded to do the merger by Tony Benn. A few weeks later Stokes told Sampson, 'If I'd known what a mess BMC was in, I'd never have agreed.'
13. Hugh Scullion, 'The skilled revolt against general unionism: the case of the BL Toolroom Committee' in the *Industrial Relations Journal*, May/June 1981.
14. *The Economist*, 19 February 1977.
15. Michael Edwardes, *Back from the Brink* (London: Collins, 1983) p.77.
16. Ibid., p.78.
17. Ibid., p.79.
18. Ibid., p.224.
19. Sampson, op. cit., p.356.
20. Edwardes, op. cit., p.117.
21. *The Times*, 20 November 1979.
22. Edwardes, op. cit., p.123.
23. Sampson, op. cit., p.356.
24. *The Economist*, 6 August 1983.
25. *The Times*, 13 July 1983.
26. Edwardes, op. cit. p.167.
27. *The Times*, 13 July 1983.

28. Edwardes, op. cit. p.289.
29. *The Times*, 30 December 1983 and 6 and 12 January 1984.

CHAPTER 11: PUBLIC SERVICES

1. C. T. B. Smith *et al.*, *Strikes in Britain*, Department of Employment Manpower Paper no 15 (London: HMSO, 1978).
2. Ibid., p.24.
3. Ibid., p.84.
4. *The Times*, 13 August 1983.
5. Conspiracy and Protection of Property Act 1875 and Electrical Supply Act 1919.
6. L. J. MacFarlane, 'Strikes Against the Government', Paper presented to the Political Studies Association Annual Conference, 1980.
7. Ibid.
8. Ibid.
9. The best book about this period is by Denis Barnes and Eileen Reid, *Governments and Trade Unions: The British Experience 1964–79*, Policy Studies Institute (London: Heinemann, 1980). Others include Gerald A. Dorfman, *Government Versus Trade Unionism in British Politics since 1968.* (London: Macmillan, 1979) and Eric Wigham, *Strikes and the Government 1813–74* (London: Macmillan, 1976). For an inside story of the Heath Government, see the book by Edward Heath's Political Secretary, Douglas Hurd, *An End to Promises* (London: Collins, 1979). For a more general survey focussing on the violent and disruptive aspects of this period see Richard Clutterbuck, *Britain in Agony* (London: Penguin, 1980).
10. Barnes and Reid, op. cit., pp.223–4.
11. *The Economist*, 13 February 1971.
12. Clutterbuck, op. cit., p.125.
13. *New Left Review*, September 1975.
14. Clutterbuck, op. cit., pp.112–13.
15. Ibid., pp.329 and 337.
16. *The Times*, 28 July 1983.
17. *The Times*, 29 July 1983.
18. For other examples of the shortcomings and hazards of anti-strike legislation, see L. J. Macfarlane, *The Right to Strike* (London: Penguin, 1981).

CHAPTER 12: THE JOHN LEWIS PARTNERSHIP AND BAXI

1. The sources for this chapter were mainly primary: a number of visits to the John Lewis Partnership central office in Oxford Street and to other stores; interviews at various levels; and attendance at a meeting of the Central Council at which the controversy arising from the Chairman's rejection of a previous recommendation was debated (as described in this chapter). Apart from these, the author has taken the John Lewis Partnership *Gazette* throughout the months of 1982–83 which cover the

greater part of the case study. The section on Baxi is based on discussions with the Chairman supplemented by press reports. The author is very grateful indeed for the time which all these people, all of whom have a heavy workload, gave up to this study.

2. These figures have been updated from the preliminary figures given in the John Lewis Partnership *Gazette*, 12 March 1983.
3. It can be argued that those with the initiative to stand are the kind of people to whom promotion will come by reason of their ability.
4 All figures taken from the *Gazette*, 9 April 1983.
5. Ibid.
6. About 5 decisions have been overruled or referred back in the past 10 years.
7. An executive of another company reading the Chairman's statement and his almost weekly replies in the letters column, expressed amazement that he did find the time but clearly he gives this a high priority.
8. *Gazette*, 4 June 1983.

CHAPTER 13: COOPERATIVES

1. *The Times*, 8 August 1983; Labour Research Department, *Fact Service*, 13 August 1983.
2. See the *Observer*, 3 January 1982 and the *Guardian*, 4 January 1982. When the Left seemed to be gaining control of the Labour Party, four of the 17 Cooperative Party MP's defected to the SDP. For over 100 years there have been two quite separate streams of British Socialism both in the Labour Party and in the trade unions. One of these emanated from the philosophies of Robert Owen and the friendly societies, and the other from the philosophies of Karl Marx, later in the 19th century. The Owenites were the forefathers of the present day 'moderate' elements in both the trade unions and the Labour Party, dedicated to getting the best possible wages and working conditions within the existing mixed economy, whereas the 'militants' were fathered by the Marxists whose aims were always political rather than industrial, to bring about revolutionary changes in society by exploiting the contradictions inherent in capitalism, and later in the mixed economy, and causing it to collapse.
3. Jenny Thornley, *Workers' Cooperatives: Jobs and Dreams* (London: Heinemann, 1982) pp. 41–45. This is an invaluable and comprehensive analysis of all the principal kinds of workers' cooperatives and of their sponsoring and supporting organizations, and this chapter owes much to her book. She also contributed to a fascinating double bill with Arthur Scargill, 'Conflict or Cooperation in Industry' in *New Society*, 7 January 1982, underlining the clash between the two streams of socialism (see note 2 above).
4. Thornley, *Workers' Cooperatives*, p.44. She records that local councils, anxious to create jobs, have been more ready to take risks but she comments '. . . in a number of cases co-operatives have been funded on criteria which are far from commercially sound and based more on political objectives'.
5. Ibid., pp.57–60.

6. Ibid., pp.33–6.
7. For an assessment of each of these and others see Thornley, op. cit., chs 2 and 3.
8. Ibid., pp.200–1.
9. Ibid., p.89.
10. Jack Eaton, '*The Basque Workers' Cooperative*' in the *Industrial Relations Journal*, Autumn 1979, p.32.
11. Peter Jay in *The Times*, 7 April 1977.
12. Eaton, op. cit., p.33.
13. Industrial Research and Information Service, *IRIS News*, May 1982.
14. Peter Jay, in *The Times*, 14 April 1977.
15. Tony Eccles, *Under New Management* (London: Pan 1981) p.68. Eccles was a Professor of Business Administration who, starting as an engineering fitter on Merseyside, won a scholarship and earned a First Class Honours degree in Mechanical Engineering, followed by 10 years in production management with Unilever. He stood as Labour candidate for Runcorn in October 1974 and was asked by Spriggs in December 1974 to advise on setting up and running the cooperative. The book gives a vivid and moving account of this and next 5 years. Like the study of APC (UK) (see Chapter 9), the author found reading his story like reading a classic tragedy in which the characters, with the best of intentions and homeric efforts, advance inexorably towards disaster. It is an exciting and salutary book.
16. Ibid., p.348.
17. Ibid., p.378.
18. Patrick Wintour, 'How not to', *New Statesman*, 5 March 1983.
19. Labour Research Department *Fact Service*, 6 November 1982 and 13 August 1983.

CHAPTER 14: MARKS AND SPENCER

1. This case study is based on visits both to the Head Office and to many Marks and Spencer Stores. A large number of executives and managers gave up a great deal of time to the author, including the (former) manager and staff of his nearest store, to all of whom he is very grateful. They prefer not to be named individually but to contribute as members of the team. These discussions were supplemented by background reading of the definitive history by Geronwy Rees, *St Michael: A History of Marks and Spencer* (London: Pan, 1973).

CHAPTER 15: ALVIS

1. This chapter is based on two visits by Anne Ollivant and one by the author to the Alvis works in Coventry and they are very grateful to the managers and union representatives who gave up their time. As with the other case studies the judgements and opinions expressed are inevitably those of the author.

CHAPTER 16: GEC TELECOMMUNICATIONS

1. This chapter is based on a series of visits to the company in 1981 and 1982 in which 6 managers and 5 trade union convenors and shop stewards (representing TASS, TGWU and AUEW) were interviewed, and a large number of other junior management, shop stewards and workers consulted more briefly in their work place. A great variety of opinions were expressed, so the interpretation is inevitably that of the author.
2. Text of Membership Agreements signed in 1969 between GEC Telecommuncations Ltd. and the Association of Scientific, Technical and managerial Staffs (ASTMS), the Association of Clerical, Technical and Supervisory Staff (ACTSS) and the Association of Professional, Executive, Clerical and Computer Staff (APEX).
3. This does not, however, seem to have been the experience in Germany.
4. The redundancy problem was also examined in another context in Chapter 9.

CHAPTER 17: JAPANESE MANAGEMENT IN THE UK

1. Toshiba Consumer Products (UK) Ltd, Plymouth, in a letter to the author, 19 November 1982. This firm, in fact, retained UK top management throughout.
2. Geoffrey Bownas, in a lecture to a conference of the Management Centre, Europe, in London, 15 December 1980.
3. Michael White and Malcolm Trevor, *Under Japanese Management* (London: Heinemann, 1983) p.128.
4. Ibid., pp.44–50.
5. Ibid., p. 69.
6. *Industrial Relations Review and Report* (IRRR), August 1981.
7. *Works Management*, July 1982, p. 47.
8. Ibid., p.45.
9. *Works Management*, August 1981, p.37.
10. Ibid., p.41.

CHAPTER 18: ALTERNATIVE ROUTES TO INDUSTRIAL REVIVAL

1. G. C. Allen, *The British Disease* (London: Institute of Economic Affairs, 1979) pp.69–70.
2. Department of Employment *Gazette* cited by Ken Coates and Tony Topham in *Trade Unions in Britain* (Nottingham: Spokesman, 1980).
3. e.g. see William G. Ouchi, *Theory Z* (New York: Avon Books, 1981).
4. Quoted more fully on page 39.
5. The Times, 11 July 1983.
6. Survey by Inbucon and Employment Conditions Abroad (ECA) reported in *The Times* the BBC *Today* programme on 11 August 1983.
7. Summarized from a letter to the author from Senator John Siddons, 3

December 1982 and from the text of Australian Senate Bill No 257 of 26 November 1981.

8. *Working Together for Britain* . . . General Election Manifesto (London: SDP/ Liberal Alliance 1983) pp.10–11.
9. Allen, op. cit. pp.21–4.
10. John Garnett, *The Work Challenge*, London: The Industrial Society, 1978) pp.21 and 49.
11. This is not just a wild swipe. It is based on the author's experience of helping graduates to get jobs during 11 years teaching at university.
12. Again, this is no myth. The author found at university that there was an intense yearning amongst arts and social science students to do practical or empirical courses but their degree requirements forced them to do many more theoretical courses than many of them either wanted or needed for their future lives.
13. In Britain engineers are too often seen as 'on tap' rather than 'on top', earning half the salary of a lawyer or accountant who has become managing director over them. This is partly due to a limited outlook often found in British engineers, who scorn 'administration' in their younger days so that, if they do reach top positions they display bad *financial* management, with disastrous results. This was not so in the days of the great engineers – Telford, Stephenson, Brunel – who had a broader vision.
14. John Garnett, Director of the Industrial Society (op. cit., pp.17–21) considers that the ideal work group at operating level should not be more than 18 (the foreman's span of control) and that in the management chain, levels should be reduced by having 7 managers reporting to the next level up.
15. Peter Walker, *The Ascent of Britain* (London: Sidgwick & Jackson, 1977), p.71. See also George Goyder, *The Responsible Worker* (London: Hutchinson, 1975) and Michael Shanks, *What's Wrong with the Modern World* (London: The Bodley Head, 1978).
16. see Garnett, op. cit., pp.32–8.
17. Much of the sickness, even, is an indirect result of the general malaise and attitude towards productivity. Some of the genuine sickness arises from exasperation and mental strain from assembly line work, especially where it is noisy; and some 'sickness' merely comprises people taking up their annual entitlement of free sick time as an escape from unsatisfying work, see ibid., pp.1–2.
18. This accumulates over the years to our even greater overall shortfall of 50% *vis-à-vis* the Germans.
19. The Warwick University Industrial Relations Research Unit carried out an excellent study of the varying nature and work of shop stewards in a large manufacturing company in the Midlands. Amongst the types of shop steward, they paid particular attention to 'leaders' and 'populists'. 'Leaders' acted as representatives rather than delegates, guided by their own principles, and carried their men with them. 'Populists' acted rather as delegates and did what they thought would get them popular support. They found that, in the event, 'leaders' achieved more benefits for their constituents than 'populists'. Eric Batstone, Ian Boraston and Stephen

Frenkel, *Shop Stewards in Action* (Oxford: Basil Blackwell, 1977).

20. In the course of our case studies we came across repeated manifestations of this constructive management/union partnership – e.g. in GEC and in 'APC(UK)'. In the latter, for example, one of the Convenors, though a leading militant in the disputes described in Chapter 9, had proposed and carried through some very positive ideas mutually beneficial to the company and to his members. He had founded, with company cooperation, a 'disabled workshop' where those injured or otherwise disabled could perform suitable productive tasks from wheel chairs; and he had introduced a 'time bank' whereby a member of a production team could build up future days off (e.g. for extra holidays) by participating in maintaining the full production target by his team despite operating one or two below strength (i.e. covering the work of those who were cashing their 'time bank' balance on that day). Alternatively a man could work extra time and 'put it in the bank'. The men appreciated the arrangement as a means of occasional relief and a better lifestyle for themselves and their families, and the management cooperated, despite some accounting problems, because it helped to maintain productivity and a 'happy ship'. APC(UK), despite the disputes described, *is* a 'happy ship', as reflected by the low labour turnover and cordial working relations. Whether the good done by a convenor in his constructive activities balances the damage done by his militancy, only the Almighty (or a penetrating study of the company's accounts) can decide.
21. France, with its huge natural resources (notably food) its low population density and the flexibility provided by its numerous, low paid, temporary immigrant labour force, should be far more prosperous than it is.

CHAPTER 19: BRITAIN IN THE 1990s

1. Anthony Hyman, *The Coming of the Chip* (London: New English Library, 1980) pp.7–8.
2. Ed Goldwyn, 'Now the Chips are Down' in Tom Forester (ed.), *The Micro-electronics Revolution* (Oxford: Basil Blackwell, 1980) p.299.
3. Reported in the *Observer*, 14 November 1982.
4. Hermann Kahn, *The Coming Boom* (London: Hutchinson, 1983) p.217. Taiwan and Mexico are examples.
5. M. J. Earl, 'What Micros Mean for Managers' in Forester, op. cit., p.360.
6. Anthony Hyman, op. cit., pp.49–52.
7. Quoted by Ian Benson and John Lloyd, *New Technology and Industrial Change* (London: Kogan Page, 1983) p.43.
8. P. Sadler, 'Welcome Back to the Automation Debate' in Forester, op. cit., p.301.
9. Tom Stonier 'The Impact of Microprocessors on Employment' in ibid., p.305.
10. Goldwyn, op. cit., in Forester op. cit., p.300.
11. In the 1880s, for example, an urgent survey was carried out in the USA to tackle the crisis in distribution of the evermounting tonnages of goods being disgorged by the railways. It was calculated that the heavy horse

population would be unable to reproduce itself fast enough and a programme was instituted for massive importation of stallions and breeding mares. One of the less reverent members of the research team projected the study ahead 100 years and estimated that the breeding areas would be under five feet of dung by 1980. Then someone invented the internal combustion engine.

Today some people are alarmed about oil running out early in the 21st century. Yet already the alternative sources of energy are in sight – e.g. power by nuclear fusion which could produce energy at ten times the predicted rate of consumption, not to mention solar, wave and tidal energy, all of which could become economical if oil became scarce. Moreover, the threat of scarcity spurs invention. Is nothing unexpected going to be invented in the next 50 years?

12. Barry Jones, *Sleepers Wake: Technology and the Future of Work* (Brighton: Wheatsheaf, 1982) pp.6–7.
13. European Commission, *Employment and the New Microelectronic Technology*, Communication to the Standing Employment Committee (Brussels: European Commission, February, 1980) Annex, p.1.
14. Sadler, op. cit., in Forester op. cit., pp.295–6.
15. Reported in *The Times*, 3 November 1981.
16. Kwangtung Provincial official in discussion with the author during a visit to China, August 1975.
17. *The Economist*, 23 July 1983.
18. For a good essay on the subject of education for the future see Stonier, op. cit., in Forester, op. cit., pp.303–7.
19. Barry Jones explores this idea in op. cit;, ch. 7 and pages 242–3.
20. Ibid., pp.204–5, 251–2.
21. Ibid., pp.50–1, 240.
22. Stonier, op. cit., in Forester op cit., p.305
23. Sir Clive Sinclair, *Futures*, BBC 2, 7 October 1982.
24. J. F. Engleberger, Lecture to Young Presidents' Organization, Jasper, Alberta, 17 September 1982.
25. Ibid.
26. Shirley Williams, *Politics is for People* (London: Penguin, 1981) p.69.
27. Barry Jones, op. cit., p.209.
28. Harold Macmillan, transcript in the *Listener*, 8 February 1979.

Bibliography

ALLEN, G. C., *The British Disease* (London: Institute of Economic Affairs, 1979).

A very clear analysis of why Britain's economic performance has been so sluggish. Suggests giving more incentives for success and fewer subsidies for failure; overhauling education to meet needs; more technological education for civil servants; and more professional training for managers

BARNES, D. and REID, E., *Government and Trade Unions: the British Experience 1964–79*, Policy Studies Institute (London: Heinemann, 1980).

A masterly account of Government attempts to involve themselves in industrial relations, from Harold Wilson's *In Place of Strife* to the Wilson–Callaghan Social Contract, by one who was personally involved as a senior civil servant for much of the time.

BATSTONE, E., BORASTON, I. and FRENKEL, S., *Shop Stewards in Action* (Oxford: Basil Blackwell, 1977).

A comprehensive study of the role of the shop steward, of different types of shop steward, their motivations, their actions and the response they get. Set in a large manufacturing firm in the West Midlands. The company and the unions gave full cooperation to the research with excellent results.

BBC, *Coverage of the Industrial Situation in January and February, 1979* (London: BBC, 1980).

A clear and concise BBC examination, in retrospect, of their coverage of the 'Winter of Discontent'.

BEHARRELL, PETER, and PHILO, GREG (eds), *Trade Unions and the Media* (London: Macmillan, 1977).

Members of the Glasgow Media Group (see also below) provide here a shorter version of their indictment of the media for bias against the trade unions, which is more fully set out in *Bad News* and *More Bad News*.

BENSON, IAN and LLOYD, JOHN, *New Technology and Industrial Change* (London: Kogan Page, 1983).

This book offers an optimistic and a pessimistic assessment of the effect of new technology on society and, in particular, on employment and on the tactics for handling the change by the Labour movement and trade union.

BERRY, A. P. (ed.), *Worker Participation – The European Experience* (Coventry: Coventry and District Engineering Employers Association, 1974).

A useful analysis of industrial democracy in Belgium, Denmark, France, Germany, Ireland, Italy, Luxembourg and The Netherlands.

THE BRITISH INSTITUTE OF MANAGEMENT, *Funds Available to Employees on Strike, 1978*.

A useful report giving the various sources of funds available to strikers.

CLUTTERBUCK, RICHARD, *Britain in Agony* (London: Penguin, 1980).

An analysis of political violence in Britain from 1970–9, including violent picketing, from Saltley to the Winter of Discontent.

CLUTTERBUCK, RICHARD, *The Media and Political Violence* (London: Macmillan, 1983).

Do the media encourage or exacerbate violence in riots, strikes and terrorism? What makes them tick? How best to work with them.

COATES, K. and TOPHAM, T., *Trade Unions In Britain* (Nottingham: Spokesman, 1980).

A comprehensive description of the aims, objectives, constitutions, membership and performance of British trade unions.

COMMISSION ON INDUSTRIAL RELATIONS, *Workers Participation and Collective Bargaining in Europe* (London: HMSO, 1974).

A very useful reference document on industrial democracy and collective bargaining practices in different countries.

CPRS, *The Future of the British Car Industry* (London: HMSO, 1975).

A very detailed analysis of the dismal state of the British motor industry in 1975, in comparison with others, containing some disturbing statistics and some penetrating comments on where the trouble lies.

CREIGH, S. W., 'The Economic Cost of Strikes' in *The Industrial Relations Journal*, Spring, 1978.

An outstanding analysis by a very able civil servant at the Department of Employment. Complements the British Institute of Management study above.

CREIGH, S. W., DONALDSON, N. and HAWTHORNE, E., 'Stoppage Activity in OECD Countries' in *Employment Gazette*, November 1980.

Some very useful statistics on strike histories in industrial countries, highlighting some of the contrasts (e.g, between Britain and Germany).

DAHRENDORF, RALF, *On Britain* (London: BBC, 1982).

A German who knows Britain very well believes that she can and must adapt to change. He differentiates between 'work' (what we are required to do) and 'activity' (what we want to do) and admires the British ability to put quality of life before economic gain.

DONOVAN, LORD (Chairman), *Report of the Royal Commission on Trade Unions and Employers Associations,* Cmnd 3623 (London: HMSO, 1968).

The Report which led Wilson to float *In Place of Strife* and, to some extent, influenced the 1971 Industrial Relations Act. The Report concentrates mainly on the causes and cures for unofficial stoppages.

DORFMAN, G. A., *Government versus Trade Unionism in British Politics since 1968* (London: Macmillan, 1979).

An American analysis of the conflict of Government and Trade Unions from *In Place of Strife* to the Social Contract.

EATON, JACK, 'The Basque Workers' Cooperative' in *The Industrial Relations Journal,* Autumn, 1979.

A good, though critical, account of the Mondragon system.

ECCLES, TONY, *Under New Management* (London: Pan, 1981).

A tragic account of the failure of KME, a cooperative supported by government money and managed by the trade union convenors.

EDWARDES, MICHAEL, *Back from the Brink* (London: Collins, 1983).

A first-hand account of the stewardship of Sir Michael Edwardes at BL, describing how both financial disaster and trade union militancy were tackled.

FORESTER, TOM (ed.), *The Microelectronics Revolution* (Oxford: Basil Blackwell, 1980).

Includes some excellent articles on the impact of microelectronics on society as well as on computer technology. The book is very well balanced, with optimists, pessimists, technologists, pro- and anti-including a graduate SWP member who 'provides rank and file activists with a step-by-step guide to blacking new technology'. Most of the contributors, however, are professional and constructive and this is one of the best books available on the subject.

GARNETT, JOHN, *The Work Challenge* (London: The Industrial Society, 1978).

A concise, forceful and readable handbook on management, with particular stress on communications, by the Director of the Industrial Society.

GENNARD, JOHN, *Financing Strikers* (London: Macmillan, 1977).

An outstanding assessment, backed by detailed research and statistics, of what strikers live on and where it comes from.

GENNARD, JOHN, *The Effects of Strikes on Individual Strikers* (London: Social Science Research Council Report, May, 1978).

A shorter version of the above, based mainly on four cases studies.

GENNARD, JOHN, 'Doling it out to Strikers' in *Personnel Management*, November, 1979.

What strikers do and do not get from social security.

THE GLASGOW UNIVERSITY MEDIA GROUP, *Bad News* and *More Bad News* (London: Routledge and Kegan Paul, 1976 and 1980).

Two attacks on the media which the authors think are heavily biased against the trade unions and against socialism. Supported by content analysis and quotations from broadcasts etc.

GOYDER, GEORGE, *The Responsible Worker* (London: Hutchinson, 1975).

A thoughtful book by an experienced managing director on balancing the interests of the shareholders, the workforce, the trade unions and the consumers.

HAYWARD, JACK (ed.), *Trade Unions and Politics in Western Europe* (London: Frank Cass, 1980).

A group of articles on British, German, Swedish and Italian Trade Unions.

HEY, JOHN, *Britain in Context* (Oxford: Blackwell, 1979).

A very valuable set of tables showing the progress of the British economy in the 20th century, comparing its performance with others.

HMSO, Draft White Paper, *In Place of Strife*, Cmnd 3888 (London: HMSO, 1969).

The Wilson Government's attempt to apply the law to industrial relations following the Donovan Report. The attempt was defeated by the Trade Union sponsored MPs in the Labour Party.

HMSO, White Paper, *Trade Union Immunities*, Cmnd 8128 (London: HMSO, 1981).

White Paper heralding Norman Tebbitt's 1982 Employment Act.

HURD, DOUGLAS, *An End to Promises* (London: Collins, 1979).

An account of the Government of Edward Heath written by his Political Secretary.

HYMAN, ANTHONY, *The Coming of the Chip* (London: New English Library, 1980).

A very clear and concise outline for the layman of what the chip can and cannot do for production, communication, office work, medicine, surveillance, education, retailing and domestic life.

INSTITUTE OF ECONOMIC AFFAIRS, *Trade Unions: Public Goods or Public*

'*Bads*'? (London: Institute of Economic Affairs, 1978).
A series of articles or transcripts of speeches by twenty people including Lord Scarman, Lord Robens, Jo Grimond and Reg Prentice analysing the place of the trade unions in British society.

JAY, PETER, in *The Times,* 7 and 14 April, 1977.
A pair of excellent articles about the Mondragon Cooperative system in Spain. Concise and full of meat, generally presenting a favourable case.

JOHN LEWIS PARTNERSHIP, *Gazette.*
A weekly house journal which is a model of free speech, encouraging critical letters, usually under a pseudonym, which are answered weekly by management. Anyone can buy it and it gives proof of the health and confidence of the Partnership.

JONES, BARRY, *Sleepers, Wake!* (Brighton: Wheatsheaf, 1982).
A very interesting and wide-ranging analysis of the problems posed by microelectronics, with well documented assessments of the prospects and of various suggested solutions. Many of his ideas are original and radical, some perhaps unrealistic, but this book is stimulating and constructive and, with Forester, one of the best.

KAHN, HERMANN, *The Coming Boom* (London: Hutchinson, 1983).
A forecast of the boom coming to the USA in the 1990s by a well known 'futurologist'. Mainly a defence of Reaganomics but does also contain useful data on a wider front.

LAYARD, RICHARD, *More Jobs, Less Inflation* (London: Grant McIntyre, 1982).
A radical proposal, now part of SDP policy, to check wage inflation by a counter-inflation tax, whereby firms which give pay rises in excess of a stated norm pay a tax penalty.

MACFARLANE, L. J., *The Right to Strike* (London: Penguin, 1981).
An excellent study of the ethics and practices of the strike weapon when used to coerce governments.

MACSHANE, DENIS, *Using the Media: Worker's Handbook* (London: Pluto Press, 1979).
A guide for trade unionists on how to go about getting their case best put over in the media. Written by an experienced journalist, it also provides a useful insight into how the media operate.

MARSH, ARTHUR, *Trade Union Handbook* (London: Gower, 1979).
Primarily a factual reference book with a brief introductory analysis.

MARWICK, ARTHUR, *British Social History since 1945* (London: Penguin, 1982).
A useful volume of the Pelican Social History of Britain.

MINFORD, PATRICK, *Unemployment: Cause and Cure* (Oxford: Martin Robertson, 1983).
A proposal to cure unemployment mainly by ensuring that unemployment benefits are not more than 70% of net income, and by making continuance of benefits for the long-term unemployed conditional on accepting work on a job pool (including community work). The author is an economist who advises Mrs Thatcher's Government.

OUCHI, WILLIAM, G., *Theory Z* (New York: Avon Books, 1981).
An account of the management theory which has swept the USA. It is based on Japanese experience and concentrates on good communication. What it advises is to be found in many Japanese firms in Britain and in good firms like the John Lewis Partnership.

PEEL, JACK, *The Real Power Game: a Guide to European Industrial Relations* (Maidenhead: McGraw-Hill, 1979).
An authoritative account of the EEC structure for industrial relations and an analysis of some of the issues involved, by a lifetime British trade unionist who became Director of Industrial Relations for the EEC in Brussels.

POLLARD, S. and CROSSLEY, D. W., *The Wealth of Britain 1085–1966* (London: Batsford, 1968).
A useful historical account of the economic development of Britain, and of its social implications.

RADICE, GILES, *The Industrial Democrats* (London: G. Allen & Unwin 1978).
The case for involvement of trade unions (as distinct from directly elected workers' representatives) in the management of industry, written by an MP on the Centre Right of the Labour Party.

REES, GERONWY, *St Michael: History of Marks and Spencer* (London: Pan, 1973).
A definitive history of Marks and Spencer from the opening of the Penny Stall by Michael Marks in 1884.

SAMPSON, ANTHONY, *The Changing Anatomy of Britain* (London: Hodder and Stoughton, 1983).
The latest in his admirable series since his original *The Anatomy of Britain.* He dissects every significant part of the anatomy: all the main political parties, the civil service, the city, industry, the trade unions, the media etc.

SCARGILL, A and KAHN, P., *The Myth of Workers' Control* (Leeds and Nottingham Universities, 1980).
A pamphlet opposing participation of workers or trade union officials in management.

SCULLION, HUGH, 'The skilled revolt against general unionism: the case of the BL Toolroom Committee' in *The Industrial Relations Journal*, May/June 1981.

A very interesting account of a conflict arising from the resentment of skilled workers at seeing their differentials eroded.

SHANKS, MICHAEL, *What's Wrong with the Modern World?* (London: The Bodley Head, 1978).

A crystal clear analysis of the roots of our economic problems offering some alternative solutions but primarily aiming to provide an 'Agenda for a New Society'.

SMITH, C. T. B. *et al.*, *Strikes in Britain*, Department of Employment Manpower Paper No. 15 (London: HMSO, 1978).

A very useful set of statistics bringing out clearly which industries are the most strike prone.

SONTHEIMER, KURT, *The Government and Politics of West Germany* (London: Hutchinson, 1972).

A useful background book on the foundations of the German economic miracle.

STEWART, MICHAEL, *The Jekyll and Hyde Years: Politics and Economic Policy Since 1964* (London: Dent, 1977).

A critical analysis of the policies of three governments in the years 1964–76 (Wilson, Heath, Wilson).

THOMSON, A. W. J. and ENGLEMAN, S. R., *The Industrial Relations Act* (London: Martin Robertson, 1975).

An excellent account of what it consisted of and why it failed.

THORNLEY, PENNY, *Workers' Cooperatives: Jobs and Dreams* (London: Heinemann, 1982).

A comprehensive account of all the principle types of workers' cooperatives, how they are constructed, funded and managed.

TREVOR, MALCOLM and WHITE, MICHAEL, *Under Japanese Management* (London: Heinemann, 1983).

A very useful study showing how Japanese managers succeed by treating the British as British and not as if they were Japanese. This book should be studied by other firms.

TUC, *A Cause for Concern* (London: TUC, 1979).

Examines the media coverage of industrial disputes in January and February 1979 to find out how and why the strikers got unfavourable coverage.

TUC, *How to Handle the Media: a Guide for Trade Unionists* (London: TUC, 1979).

A more concise guide than MacShane's book (above) on how trade unionists should handle the media to get their case put over favourably.

TUC, *Programme for Recovery: TUC Economic Review 1982* (London: TUC, 1982).

A statistical analysis of the British economy and unemployment, proposing a cure consisting largely of increased public expenditure.

TURNER, H. A., *Is Britain really Strike-Prone?*, Occasional Paper no. 20 (1969), Department of Applied Economics, Cambridge, 1969.

A paper by the Professor of Industrial Relations at Cambridge suggesting that reports of strikes in the motor industry give an exaggerated impression of their real effect.

WALKER, PETER, *The Ascent of Britain* (London: Sidgwick and Jackson, 1977).

A readable exposition of mainstream or moderate Conservative policies for tackling Britain's economic problems by one who has been a Minister in both the Heath and Thatcher Governments.

WIGHAM, ERIC, *Strikes and the Government 1893–1974* (London: Macmillan,, 1976).

An excellent analysis of the period by an experienced labour relations journalist who was a member of the Donovan Commission in 1965–8.

WILLIAMS, SHIRLEY, *Politics is for People* (London: Penguin, 1981).

A plan for the future by one of the founders of the Social Democrat Party, including some valuable thoughts on adjusting to new technology.

WINTOUR, PATRICK, 'How not to'. *New Statesman* 5 March 1983.

A review of Tony Eccles' book on the KME cooperative (see above) which gives a succinct assessment of why it failed.

Index